Musings

STORIES OF
BYGONE DAYS, POEMS, AND
INSPIRATIONAL THOUGHTS

by
DONALD MANKIN

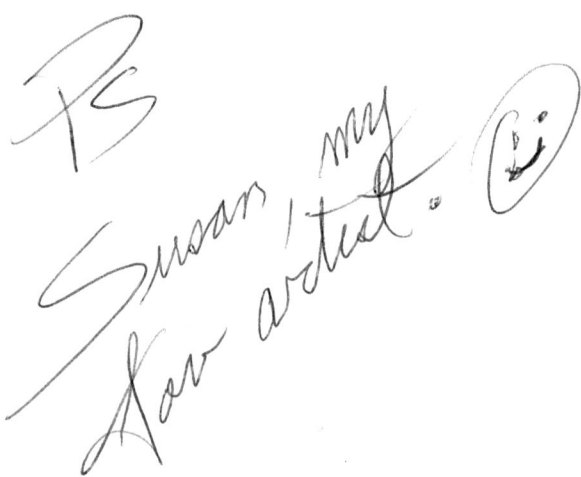

The author acknowledges the trademarked status and trademark owners of various products referenced in this work of nonfiction, which have been used without permission. The publication/use of these trademarks is not authorized, associated with, or sponsored by the trademark owners.

Musings: *Stories of Bygone Days, Poems, and Inspirational Thoughts*

Copyright © 2018 by Donald Mankin
Cover Design by Lori Jackson
Editing by Steede Mankin

All rights reserved. Without limiting the rights under copyright reserved above, no part of this publication may be reproduced, stored in or introduced into a retrieval system, or transmitted, in any form, or by any means (electronic, mechanical, photocopying, recording, or otherwise) without the prior written permission of the above copyright owner of this book.

Library of Congress Cataloging-in-Publication Data

Mankin, Donald
Musings: *Stories of Bygone Days, Poems, and Inspirational Thoughts* — 1st edition
ISBN-13: 978-1976806056

To Jody

If I could walk among the stars
And Heaven's wonders there unveil,
I would cut off Halley's Comet's tail
And bring it back to Jody.

If I could stroll the sea's floor bold,
A Spanish galleon full of gold
I would bring back home to Jody.

But I can't walk up in the sky
Nor down beneath the sea,
So, galaxy and gold may pass her by
But never, never me.

Table of Contents

	Acknowledgments	i
1	Christmases I Remember	1
2	God Art	6
3	Ladies Should Not Sit Along the Wall	7
4	Ask	8
5	Nine Wrinkles and Four Chins	10
6	Freedom, What's It Worth?	12
7	Daughters	14
8	Along Came the God Particle	15
9	Coffee Sure Tasted Different Back Then	16
10	Country Boys and Fedora Hats	18
11	Hope	20
12	Imitators of God	22
13	A Matter of Perspective	25
14	Mothers, Thunderstorms, and Crying Little Boys	27
15	Depravity of Man	30
16	At the Altar of Mammon	32
17	They Died So Young	34
18	Coffee, The Right Way	36
19	From Hangnails to Hangings	37
20	Big John and Me	39

21	The Greatest of Conflicts	40
22	Of Fireflies and Fruit Jar Lanterns	42
23	Who Is Righteous?	44
24	Sound the Alarm	47
25	For the Unborn	48
26	Oh, The Lion Roars Tonight	50
27	Remembering Snip	53
28	Growing Up Okie	55
29	Of Drylines and Texas Thunderstorms	58
30	Praise Jesus	60
31	Another Visit to the U.S. Virgin Islands	61
32	Of Screen Doors, Latches, And Prankster Sisters	63
33	Do Not Get Cozy with a Macaw Parrot	65
34	Eventually Time Leaves Everyone Behind	68
35	Not Really	71
36	Sannado	74
37	Hound Sounds	78
38	Of Crocheting, Quilting, Knitting, and The Like	80
39	Candles, Kerosene Lamps, and Electricity	81
40	Where Panthers Scream and Mad Dogs Roam	83
41	Lost in Peachtree Hollow	87

42	For Nadine	90
43	Long, Long Road to Alaska	93
44	Herman's Worthless Dogs	110
45	Henry's Headless Turtle	115
46	Wash Day in Rural Oklahoma in the Nineteen-Forties	118
47	El Comandante	121
48	Always Be Gentle	125
49	If You Ain't a Backee Chewer, Leave the Beechnut Alone	127
50	Madchen, Madchen, Madchen	131
51	Stay Between the Lines, But Which Lines?	133
52	Roasted Chicken Tasted Better in Seventy-One	135
53	On My Way to China	137
54	Time in a Box	138
55	Angels Who Guard	140
56	Now That, Friend, Was A Real Hanging	142
57	Hello, Momma, I'm Gene Poag Jr, and I'm Here to Sing a Song for You	144
58	Lay That Pistol Down Babe	146
59	Never Bear Hunt with a Three-Eighty	150
60	Stop the Bus, We're Getting Off	153
61	Don't Try to Make Little Bear by Bedtime	155

62	If You See Eyes Under the Hen, Come Back Later, Much Later	157
63	Civilian Once More	160
64	ALERT, Prankster on the Loose at Amon Carter Field	162
65	Alert! Prankster loose at Amon Carter Field ... Part II	165
66	Where There is Smoke, Look for John	167
67	I Can't Think About Jesus	170
68	Momma, I'm Cold	173
69	Of Thongs and Things	176
70	If You Eat At La Madeleine's, Learn French	178
71	Don't Stare Too Long at a Barracuda	180
72	Hidden Sins	183
73	Wilburton Main Street on Saturday Night	185
74	"Holy Horrors to Hell"	187
75	Where Is the Runway?	189
76	God, Ice Cream, and Watermelons	192
77	Crowley, A Town That Kept Us Alive	194
78	Pride in the Red, White, and Blue	196
79	Of Flashlights, Blankets, Books, and Far-Away Places	198
80	It Was A Mighty Good Roast	200
81	Walk in the Light	201
82	Khan, A Dog Worth Remembering	202

83	Thanksgiving Day Two-Thousand-Seventeen	205
84	Henry was a Prankster, Too	208
85	Nightmare at Midday	210
86	Let There Be Light	211
87	No Siree Bob, Rudolph Ain't No Sexist	213
88	The Longest Time on Death Row	215
89	In A House of Eleven, Old Hens Did Not Have Long to Live	217
90	Past and Future	220
91	The Fire	221
92	She Could Have Been from Royal Stock	223
93	Tornado	224
94	The Taste of Texas	225
95	A Bridge	226
96	Winding Road	227
97	Jody's Owl	228
98	Moments	229
99	Friends	230
100	Candles, Kerosene Lamps, and Electricity	231
101	Country Boys and Girls Know About Things They Do Not Even Know They Know	233
102	One, Two, and Many	236

103	Prom Dresses in The Sky	238
104	Henry and The Talking Deer that Threw Rocks	240
105	The Five-Year-Old Rancher	243
106	Peace That Passes Understanding	248
107	The Days of Our Years	250
108	Sound Systems Then and Now	251
109	Five Nails That Turned the World Upside Down	253
110	It Was Shelter	255
111	The Light of Christ	257
112	Empty Buckets of The Hearty	258
113	The Old Home-Place	260
	A Final Word	263

Acknowledgements

I wish, first, to express my deepest gratitude to my wife, Rheta. She never stopped encouraging me to put my stories and thoughts in a book and often jogged my memory about things we did together. Thanks to Steede Mankin for editing my manuscript, and Michelle Mankin and Page Cole for giving advice on publishing the book. I appreciate the many friends who read my stories on Facebook and urged me to publish them.

Christmases I Remember

A typical Christmas in the Mankin home during the nineteen-forties went something like this: within a day or two after Thanksgiving, Momma got a saw from Dad's toolbox and headed for the field. We children trudged along behind, sometimes in the snow, as excited as if we had been turned loose in a candy store. We lived on three-hundred-plus acres, so there was no shortage of cedar trees from which to choose. She always found a shapely one to cut down and drag to the house. The rest was left to us children.

Under Momma's tutelage, we made decorative ropes by stringing red berries picked from bushes that grew wild on our place. Dad always grew a row or two of corn. Momma popped the corn and we strung it. When we completed the stringing process, we used our handy-work, along with narrow store-bought strips of foil called icicles, to decorate the tree. We hung multi-colored glass ornaments and gold and silver colored ropes that Momma bought at Woolworth's Five and Ten Cent Store. We did not have electric lights to hang until I was eleven-years-old. That is when we got electricity, thanks to the REA (Rural Electrification Association).

While we decorated the tree, Momma made popcorn balls by combining sugar, sorghum molasses, a measure of butter, baking soda, vanilla flavoring, and salt. She brought the mixture to a boil and carefully watched until it reached a point where a soft ball formed when dropped into a glass of water just as one does when making fudge (which she always did at Christmas time).

After letting the mixture cool, she shaped it into balls and dunked them into a bowl of popcorn. Of course, the popcorn tasted wonderful, but what lay below was the real delight. I think, perhaps it was the popcorn balls we desired almost as much as the presents under the tree.

During the Christmas season, bowls of nuts (pecans, English walnuts, Brazil nuts, hazel nuts and the like) could be found on our dining room table. I can taste them today. I feel the warmth of the fireplace around which we all gathered. Listen. Do you hear the cracking of the nuts? Can you see the flames flare as we toss the shells into the fire? Still, today, I miss that tradition. There is something about cracking a pecan between one's teeth that has no equal. Of course, my dentist does not allow me to do that now. Today, we buy them already shelled. I think I might just ignore my dentist's advice next Christmas, if God lets me live until then. Just one more time I want to experience feeling the pecan shell collapse under the pressure of my bite. Once more, I shall toss the empty shells into the fire and watch them flare up as I enjoy their sweet meat. Now that sounds better than picking them out of a bag you bought at the store, right?

Of course, Decembers in Oklahoma were always cold and very often we had several inches of snow on the ground. This made for a happy occasion. Who would not be ecstatic about a white Christmas? Bing Crosby was one of the most popular singers back then and his recording "White Christmas" was played over and over on most radio station throughout the nation. Crosby sang "White Christmas" and it became synonymous with Christmas. Oh, what fun Christmases were back then!

We had several black walnut trees on our land and we gathered the nuts in a "tow sack" (burlap bag). Harvesting the nuts involved placing a walnut on a rock and smashing it with another rock so that the meat of it could be picked out with pins which women used to keep their hair in place. We did not mind that chore at all because Momma used the nuts to make a delicious black walnut cake for us to enjoy on Christmas day, after consuming a table full of delightful dishes offered by a mother who cooked with love. She always prepared tasty meals, but on Christmas day she really shined.

One year, my brother, A C, brought home a Canada Goose that he had killed. It was a Christmas that really stands out as one of the best. That was a long time ago and was my first and last goose to eat. Of course, we always had fried chicken and sometimes rabbits and squirrels that dad brought home. My memory fails me as to whether it was at Christmas time or another time, but once dad brought several quail home. He was driving his Model A Ford down the road that led to our house when a covey of quail flushed and flew across the road right in front of him. Several were killed. That evening we had those tasty little morsels for supper.

Each Christmas Eve when we were small, my youngest sister and I were hustled off to bed soon after supper. Our house had single walls so there was no insulation and the bedroom side of the house got very cold at night in the winter. We got into our warm pajamas, long-johns for me, and kneeling with our mom at the side of the bed we said our prayers and jumped into the middle of a lush feather bed. Momma always kept the feathers from chickens she killed. In our house, there were eleven, counting parents and children, so we ate a lot of chicken. Feather beds were the best part of night back then. I sunk into the feathers and found refuge there from not only the cold night air but also from any monsters that might be lurking in the dark.

But on Christmas Eve, sleep came late. I fought it. I wanted to be awake when Santa left presents under the tree. Ah but, "He knows when you are sleeping and knows when you're awake," so I never, ever caught even a glimpse of the bearded, benevolent, boisterous, old gentleman. But, I knew that he had come because the evidence was under the tree the next morning.

After opening gifts, Momma read to us from the large family bible that covered her lap. She read the Christmas story from the book of Luke. I smile as I think about how much of the story we absorbed. After all, there were toys beckoning, "Come play with me." Interestingly, it seems, even with limited financial resources, Momma always wisely chose because I can't remember ever being disappointed with my gifts.

Among my most memorable Christmases is the year that I found a green wind-up turtle under the tree. I had so much fun playing with it. Another great year was when Santa brought me an arcade. The Mario Brothers had nothing on this thing. It came with two spring-loaded, rubber sticker guns. Wind it up. Push "Start" and then shoot at a row of ducks advancing across the stage. Excellent!

One year I got a red tricycle. Man, oh man! I wore ruts in the yard riding around and around the house. But I suppose the absolute best Christmas with respect to gifts received was the year my sister and I found two J.C. Higgins bicycles standing near the tree. Hers was blue. Mine was red. Dad took us out to the driveway and demonstrated how to ride. I never even suspected he could ride a bike. As I think about it, he did sort of wobble around a bit. You know, I think maybe he just wanted to ride my bike.

Many Christmases have come and gone and my perspective of them has markedly changed. For me now, as is so often heard at Christmas time, "Jesus is the reason for the season." Christ, born to a virgin, God incarnate, in His thirty some-odd years laid out a plan to save us from our sins. The plan, "Believe on the name of the Lord Jesus Christ and you shall be saved," unfolded in eternity past.

The bible tells us that when the time was right, God sent His only begotten son, Jesus, into the world to die a substitutionary death in order that we, in believing on His name, might have eternal life. If we believe that Jesus is the son of God and believe that He died on a cruel cross to pay for all our sins, then we will be saved.

Forget the gifts under the tree. Jesus has given me the greatest gift of all. He gave me Himself. There is no greater gift. I can't have a greater gift than this: Christ died to pay for my sin.

God Art

Today, I bought a pair of pants and a shirt. Rheta usually is there to tell me if she approves. I asked the lady if they matched. "You don't have to match colors. You just make sure the colors blend, and these do. You did a good job."

So, at home we were watching a television program called "Nat Geo Wild," and saw a beautifully colored bird. It got me thinking. Have you ever seen a bird with clashing colors? I have not. Do you know why? I do. It is because God does not make mistakes.

Ladies Should Not Sit Along the Wall

In the early seventies, when I worked as a Hurricane Preparedness Meteorologist for the Caribbean, we lived in San Juan, Puerto Rico. We were close friends with a couple who were native Puerto Ricans. He was a reserve Navy officer and had privileges at the officer's club. We frequently joined them for dinner there, and on our way home we stopped at one of the casinos where I played blackjack. I usually managed to win enough to pay for our Chateaubriand dinner.

On our first such outing, Rheta took a seat on a bench that extended along the wall near the blackjack tables. However, her relaxation there was short lived because Maria, our Puerto Rican friend, came over and informed her, "Rheta, you can't sit there. This is where *las damas de la noche* (ladies of the night) sit."

Ah ha! That explains why so many men were gathered there.

Ask

What do you think of when you see the word "ask"? Is it only an action verb to you? Oh, but it is so very much more. ASK, when viewed as an acronym, contains a deep theological concept, God's perfect love for His imperfect creatures. When you see the word, as used by Jesus in His sermon on the mount, you will see that it contains power. Indeed, it contains God's full power. In it is embodied the whole of the gospel. For the person who has never come to the point in life where they see themselves as totally depraved when compared to a righteous God, the word points the way to salvation. Ask, Seek, Knock...ASK. Ask how you can be saved. Seek the only one who can save you and knock on His door. The promise from God is, "Ask and you shall receive. Seek and you shall find. Knock and it shall be opened unto you."

The embedded gospel is this: because of the original sin of Adam and Eve (they disobeyed God), every person is born into the world with a fatal condition that theologians call imputed sin. This sin nature was inherited from Adam. It is a condition that can't be removed by any means other than the blood of Christ. You can't wish it away. You can't lay claim to your parents' relationship with God. You can't work it away by doing good. "There is none righteous, no not one." Romans 3:10 "All have sinned and come short of the glory of God." Romans 3:23. So then, we ARE sinners and "The wages of sin is death, but the gift of God is eternal life through Jesus Christ our Lord." Romans 6:23.

You might ask, how can I obtain this gift of eternal life? Simply, "…if you confess with your mouth that Jesus is Lord and believe in your heart that God raised Him from the dead you will be saved. For, with the heart you believe and are justified and with the mouth you confess and are saved." Romans 10:9-10. If you truly believe in Jesus and confess Him to others as Lord, he will save you.

Did you do this at one time in your life? If you did, congratulations! You are a child of God. Ask God to save you, seek Him with all your heart, and knock on His door. ASK. He will save you. ASK is also applicable to believers. By asking, according to the will of God, you will experience all His power to live your life according to His will. God bless America and return our nation to the center of His will.

Nine Wrinkles and Four Chins

Yesterday, I changed my profile picture on Facebook. When I opened Facebook, a dear friend of mine, whom I shall just call Bill, had posted a comment on my profile pic.
"Don, what's that thing hanging down below your chin?"
Bill thought he was referring to my double chin. He was wrong. Since I have lost almost thirty pounds, what he thought was a double chin was just an empty sack where my second chin used to be. To get rid of it, I would have to have surgery. I am not going to do that because one day I might start eating again and then I would need to put my double chin there. Bill's comment brought back a memory of which I am very fond; about a young lad, quite some time ago, in a classroom for emotionally disturbed children.

I had retired from the National Weather Service and gotten very bored after six months, so I sought relief from boredom by taking a substitute teaching job.

My second assignment was at an elementary school in East Fort Worth. The school's focus was on special education and, more specifically, education of emotionally disturbed children. On that day, I was assigned to work one-on-one with a student— a fourth grader. We'll call him Jimmy (not his real name).

Back then it was still acceptable practice to isolate students. Jimmy had gotten out of control and I placed him in a time-out room. We could not lock the doors, but we could sit with our chair blocking the door. As I sat in front of the door to the room where I had placed Jimmy, I listened to all the banging, and cursing, and kicking, and stomping, and other things that are not mentionable.

After some negotiating with Jimmy, I agreed to let him out if he would behave. He promised he would. As I opened the door to free Jimmy from captivity, he blasted his way through the door like a Trident missile and headed down a long hallway toward an exit door. Because chasing students was not an uncommon task back then, I had worn my tennis shoes. So, I took off after Jimmy like a scalded dog and, just before he got to the door, I managed to get close enough to reach out and grab his shirt.

As I, and another teacher, carried Jimmy kicking and screaming back down to the time-out room, he added a few new words to our vocabulary. When I closed the door on him, he responded with, "I hate you, you wrinkled old man. I counted nine wrinkles and four chins."

My reply to Jimmy was, "I may be a wrinkled old man, but I caught you, didn't I?"

He laughed and then he was okay the rest of the day. He had become a friend to that old man with nine wrinkles and four chins.

Freedom, What's It Worth?

When I think of individual rights, I think of many of the rights enumerated by John Locke. However, whereas Locke referred to them as "natural rights," I prefer to call them "God given rights," since natural rights imply they were given to us by "Mother Nature" which is, no doubt, an affront to Almighty God. Locke was right, though, to call them inalienable rights, which could not be taken away by governments. And, it is still true that no government can take them from us.

"Well then," you might ask, "how is it that we have lost so many of our rights?"

Easy answer. They were not taken from us by the government. We gave them away. We gave them away in return for favors from our government. We forfeited them because we acquiesced to lawmakers who convinced us that we work for the government when, in fact, it is the other way around. They work for us.

We are told that we must get permission to protest peaceably in our cities when "right and good" is on our side. Yet, we watch anarchists violently take over peaceful rallies with impunity. Clamoring for "equal rights" is heard everywhere but apparently only applies to those who would deny those rights to those who disagree with them.

Each time we take something the government offers us, we give away a measure of our freedoms. I saw a picture posted on Facebook of a man burning our flag. The caption said that burning our flag should be a felony. I disagree. If we pass a law to make it a felony to desecrate our flag, then we will have surrendered our God-given right to freedom of speech.

I hate the treatment our flag gets by those who hate America, but—as one who believes in Jesus—I am to love my enemy. So, I must guard against unrighteous anger for the bible says, "to be angry and sin not." Ephesians 4:26. The man burning the flag has the Constitution behind him. In the Revolutionary War and all the wars since, individuals have fought and died to both establish this nation and to keep it free. We must not ignore the restraints of our United States Constitution.

The blood spilled over the grounds of eastern North America, Gibraltar, British India, the waters of Caribbean Sea, and the Atlantic in the war for independence demands it. The sacrifices of parents who lost a child to the ravishes of war demand it. The fallen warriors and those who lived to return home after World War I, World War II, the Korean conflict, Vietnam, and the war on terror demand it.

Will we let the things that bother us, yet are not contrary to our constitution, cause us to amend the most important document we have apart from the bible out of which the principles upon which our nation was founded sprang? Freedom. What is it worth? Not much to those who burned the flag and rioted in San Diego, and to those who want the government to rule their lives and dole out pittances to them. But, to those of us who love the United States of America, I say it is worth everything we paid in blood, and treasure, and more. Freedom. What is it worth to you?

Daughters

A couple of days ago, I went into a Mexican restaurant where I have been several times with Carrie, my daughter. The lady from whom I ordered my lunch asked, "Where is your wife?"

Rheta does not like Mexican food much, so I was puzzled. After I got home I made the connection. The server thought Carrie was my wife.

Oh, I felt like I was in my forties for a while, until Rheta suggested, "She probably thought, 'Look at that old geezer with his young wife.'"

I must talk to that girl about bursting my bubble of excitement.

Along Came the God Particle

For decades, scientists have known that the atom is not the smallest particle of matter. The standard model of particle physics has long predicted sub-atomic particles. Photons, neutrons, and quarks are some examples of these.

Then along came the "Higgs boson," also referred to as "the God particle." It is so small that its existence can only be conjectured by observing how it effects larger particles. Scientists in Switzerland are trying to confirm the existence of the Higgs boson particle using the Large Hadron Collider, a particle accelerator. If their efforts are successful, we can be assured of one thing; God created the "God particle" before Peter Higgs—the atheist for whom it is named—was born.

I will praise the One who created all things seen and unseen.

Coffee Sure Tasted Different Back Then

As I measured out a half-pound of Guatemalan Huehuetenango green coffee beans and poured them into my small home coffee roaster, my mind suddenly went back to long ago. I was back on the Poteau River that runs through Summerfield, Oklahoma.

At any rate, what does coffee have to do with Summerfield and the river that "runs through it?" It has a lot to do with reminding me of my love of coffee and my childhood memories of coffee brewed on the banks of Poteau River.

As a child, I developed a liking for the taste of coffee when my mom would give me a little of it with a greater proportion of milk. She would get up early in the morning and build a fire in the old wood burning cook stove. Then, she placed a coffee pot—a percolator as I recall—on the hot surface. After some time, the wonderful aroma of coffee brewing would waft its way to my olfactory system. The coffee was always either Folgers or Maxwell House. As time passed, she allowed me to have the real stuff. I liked it. No, I loved it.

I have been a connoisseur of coffee ever since. Connoisseur? Isn't that a mighty big word for a country boy? Maybe. I reckon it means a special knack for judging things like music or food or, in this case, coffee. It could be French, but, then again, it might be Russian or German or Greek. I wouldn't know, but when I grow up I want to be a language connoisseur. Oh wait! I am grown up, so I best get back to my story.

Some of my fondest memories of coffee were the times

when I went on two or three-day camping and fishing trips with my dad and his friends. Henry, the storyteller, would rise early and brew a pot of coffee for the rest of us. He was largely the reason I derived so much pleasure from those trips. He had a tent that could sleep six, and Henry could brew a mighty fine pot of coffee. I watched as he would drive two forked limbs into the ground and lay another smaller diameter limb across those, forming a bucket hanger. Then he would build a fire under the bucket. Once the water came to a rolling boil, Henry would toss in a handful or two of ground coffee. Since the coffee tended to boil over, Henry always had a green twig to lay across the bucket. Magically, the coffee would settle down. All the times he made coffee I do not remember once that it boiled over.

I think perhaps the thing that made the coffee perfect in every way was the fact that while the pot boiled, and the coffee came to the right strength, Henry kept us entertained with stories. He was a storyteller extraordinaire. He would take a little bit of truth and embellish it with generous amounts of imagination. Always, the result was a wonderfully told tale. A couple of my favorites, as I recall, were "The Headless Turtle" and the "Talking Deer That Threw Rocks."

I was always delighted to have an invitation from my dad to go on those outings. I have not had coffee made in a bucket over an open fire since those days. I wonder if coffee made that way today would taste as good as I remember. I think not. I believe I will just stick to my fresh roasted Guatemalan Huehuetenango and perhaps some Colombian Supremo, and dream of my childhood days with my dad, Henry, and Uncle Press on the river bank.

Country Boys and Fedora Hats

The *New York Times* carried a piece recently announcing that the mayor of Paris, France was going to implement a project that would cut car traffic along the banks of the Seine River.

"Joy to the world," I shouted as I remembered my near tragedy there.

The weather was beautiful the day we (Rheta, my son, his family, and I) stepped out of our hotel about six or seven blocks from the Seine River on our way to see the Eiffel Tower, but the wind was quite brisk.

I was never one to be too interested in hats. Oh, I wore straw hats when I mowed yards in Panola, Oklahoma. I also wore the warm kind in winter, the ones with flaps that you pull down over your ears so that I looked like "Ned and the First Reader." But I'm talking real hats here, nice ones.

Only hours before, I had found a very great looking Fedora hat in a small Paris shop. I couldn't resist. I dug deep into my pocket and managed to pull out enough Euros to pay for it. I then made a beeline for a mirror where I stood and admired myself until Rheta tugged at my arm.

"Don, come on. They are leaving without us and you're starting to embarrass me."

So, I proudly said goodbye to the shop owner and followed my family to our hotel where we napped for a bit and then headed for the Seine.

The moment we stepped out into the street I knew I must

guard my new headpiece with my life, if need be. It almost turned out that way. I was not about to lose such an eloquent head cover, so, when the hat blew from my head, I took off after it like a greyhound chasing a rabbit.

Time and again, in the six-block distance, I got oh-so-close to being able to reach down and grab it. At times, it would stop and taunt me, letting me believe I had finally won. Then, just as I would reach for it, the wind would gust again and off it would go like a Texas tumbleweed on a hot summer day. Was God teasing me?

I knew I had lost track of reality when I heard Rheta call out, "Don, you are about to run out into the expressway."

Man, oh man! Never was such a road seen in Panola. This road was wide! My heart leaped up into my throat as my beautiful Fedora skip-bounced out into that expressway and was flattened like a pancake made by a French chef.

As I watched, a deep flood of depression came over me. Then, I was encouraged by the thought that, since the hat was now flat, it could not do any more rolling. Flat things do not roll.

So, just as I was about to weave and dodge my way to the hat, a taxi driver saw it. He backed up, opened his door, and grabbed my Fedora. Alas, I would never see it again. I was going down to the Seine and drown myself.

Just before I did, though, I heard my son say, "Dad! Don't do it. I'll buy you another one. After all, it only cost ten dollars."

Hope

"Hope springs eternal in the human breast;
Man never is, but always to be blessed:
The soul, uneasy and confined from home,
Rests and expatiates in life to come."

~ Alexander Pope, *An Essay on Man*

Anarchist run rampant in our city streets and no one punishes them. Our governments and corporations are clamoring for boys and girls, men and women to use the same public restrooms. Crime is soaring in cities like Chicago where citizens are punished for defending themselves, yet the bad guys are given a free pass. So-called protesters destroy public and private property. Politicians collude to thwart the will of the people. Our Supreme Court continues to ignore the Constitution, which they swore to uphold. Our dollar is on the verge of collapse. America will be the next Greece. And on, and on, and on.

Oh, where is our hope? In whom does it lie? Well, Pope got it half right, I think. It does lie within the human breast. However, such is not true for every human breast. But it is true for those who have placed their trust in Jesus Christ. A close and truly objective look at the world's great religions will show that, indeed, speaking of Jesus, "...there is salvation in no other, for there is no other name under heaven that has been given among men by which we must be saved." Acts 4:12. From every religion there can be found something that sounds good when pulled out of context. But the truth is, a false hope is resident in dead men whose bones still can be found in their tombs. And, furthermore, for you to get to heaven or some higher form of life you must earn your way there by doing works, observing rituals, or somehow achieving a better status in this life.

Christianity alone is the exception. It is not a religion but rather a relationship. It is a relationship with One whose bones can't be found on this earth anywhere because, "He is not here. He is risen as He said! Come and see the place where He was lying." Matthew 28:6.

This relationship is possible only because God sought man rather than man having sought God. John 6:44 relates the words of Jesus, "No one can come to me unless the Father who sent me draws him." Jesus Christ, God incarnate, who was resurrected from the dead after He offered himself on the cross to pay for your sins and mine is the only hope for this world and its inhabitants.

To have a saving relationship with Jesus you must "...confess with your mouth that Jesus is Lord and believe in your heart that God raised Him from the dead..." Then you will be saved. Your sins, past, present, and future will have been paid for. You only need to accept the free gift from God. If you do not turn from sin and receive the gift of salvation from God, then you have no hope. You will pass from this world into an eternity in hell.

To discuss hell is not my intent here. Whatever your views of hell are, it is a fact that nothing good will be there. You will be eternally separated from the only one who can give you good things. The good things you enjoy here on earth are not things that you acquired on your own. Even though you might hate God, He extends to you "common grace" from time to time so that you might see His goodness and turn to Him. Romans 2:4. Those of us who believe in Jesus "...shall not perish but will have everlasting life." John 3:16. So then, for believers, Pope's words are true. "Hope springs eternal in the human breast.

Imitators of God

"Therefore, be imitators of God, as beloved children; and walk in love..." Ephesians 5:1.

Oh, what a great demand the apostle Paul placed on us who are believers in Jesus Christ! Well, Paul did not actually require us to be imitators of God. God did. He just used Paul to pen it for us. Nevertheless, when one considers the world in which we live, isn't it impossible to imitate God?

Well, yes, it is if you are not in Jesus Christ.

Then how can one be in Christ, and what does it mean to be in Him?

How? Simple. Grab ahold of the gospel.

But what is the gospel?

Well, the whole of it is embodied in one man, a God man, Jesus Christ.

The message of the gospel is this; Jesus saves.

Saves? From what?

He saves us from our slavery to sin.

What? We are sinners?

Yes. God created Adam and gave him a wife, Eve. He gave them a beautiful place of eternal abode, the garden of Eden. He gave them everything in the garden except one tree, the tree of knowledge of good and evil. Genesis 2:16-17. God gave Adam and Eve a choice to obey or disobey. They chose to disobey by eating the forbidden fruit and, in doing so, brought upon themselves a curse that has followed all of mankind down through the ages. The sin and curse are passed on to every newborn in a process that theologians call "imputed sin." Romans 3:23 says, "all have sinned..." and further that, "there is none righteous..." Romans 3:10. But the most dreadful thing about our condition is this; "the wages of sin is death..." Romans 6:23. Because of imputed sin we are born to die, and after death comes judgment. Hebrews 9:27.

So, is there no hope?

Oh yes, praise the name of Jesus who gave us our hope. "...the gift of God is eternal life through Jesus Christ our Lord." Romans 6:23.

How do we obtain this eternal life?

We simply confess that Jesus is Lord and believe in our hearts that God raised Him from the dead. Romans 10:9-10. The entirety of the Old Testament is an unfolding of the gospel that culminates in this prescription for our being freed from the chains of sin that bind us. It is laid out for us more clearly in the New Testament. Does that mean we will not ever sin after we are saved? Oh no!!! We will sin against God until we reach a perfectly righteous state. But, that will not be until we stand with Jesus in heaven.

What it does mean, however, is that when we do sin, we will not be condemned to hell. God may chastise us to bring us back when we begin to distance ourselves from Him because of having yielded to temptations to sin, but He will not forsake us because we did sin. We are His children now (1 John 3:2) but once were His enemies. Colossians 1:21. God will not forsake his children because "...He understands our frame and remembers that we are dust." Psalm 103:14. Now, having become His children and being guided by His word—the holy scriptures—we must imitate Him. As was stated at the beginning. We must walk in love.

Every decision we make, every uncertainty we face, must be evaluated in the light of Christ. What would Jesus do? When we walk in the light of Jesus, our whole perspective changes and so do our priorities. Worship of God takes the front row seat in our busy lives. Praise of Him and a lifestyle that glorifies the name of Jesus becomes foremost in our daily walk through life. We stumble often, but we will always reach up for His ever-present hand for help in times of temptation, discouragement, and trials. Caring for others and sharing the gospel with them will become more important than caring for ourselves. When we walk in love, God will use us to meet the needs of those who truly can't help themselves.

If you want to test the truth of the scriptures regarding helping others, then I challenge you to ask God to make opportunities for you to help those in need; whether it be sharing the gospel with non-believers or meeting physical needs of people. My prayer is that all who believe in Jesus will grow closer to Him daily and that we WILL walk in love so that the good news that Jesus saves will spread among our circle of friends and acquaintances and likewise to each one in their circles. May we "walk in love" and, in doing so, glorify the name of Jesus.

A Matter of Perspective

I do not remember exactly when it was, but I was probably about six or seven-years-old, which would have put it about nineteen-forty-three or nineteen-forty-four. I remember it as clearly as if it were yesterday.

We were standing in the front yard, two of my sisters and me. I can still hear their words as we looked up at the formation of airplanes flying overhead. "Those are bombers."

At that observation, I panicked. I remember the fear that gripped my young heart. I did not know a whole lot about the war, but I knew enough to make me fearful. My brother and brother-in-law were fighting their way through Italy about that time and my sister in law, Rena, kept a scrapbook of news about the war.

I was relieved to hear one of my sisters say, "They're our bombers."

Why did that memory surface? I think it did so because I had been preoccupied with thoughts of the great uncertainty about the health of our nation. I guess the question that triggered the memory was, "Have things ever in our nation's history been so dire?"

Well, yes, they have. World War II was fought over ideals much like we face today, evil men desiring to conquer us because they sought for power. Only, now those men are our own. They want to confiscate our guns and tell us the government grants us our rights instead of the Almighty God.

Then, I remembered. "We walk by faith and not by sight." 2

Cor 5:7. Of course we do. As believers in Jesus Christ, we know that "all things work together for good to those who love the Lord and are called according to His purpose." Romans 8:28. What does that mean? Well, it means that if a despot president takes away all our rights and we find ourselves being persecuted, our sovereign God will cause it all to work for our eventual good. Yes, it even means that if a formation of bombers flies overhead dropping bombs on American cities it will all work for our good. How can that be? I do not know, but I walk by faith and not by sight. So, I believe it. I rest in the knowledge that my Heavenly Father will work all things for my good.

Mothers, Thunderstorms, and Crying Little Boys

As potentially severe storms approached the metroplex from the west, a memory surfaced from long ago in eastern Oklahoma. As I gave that memory free rein to run where and as far as it chose, I suddenly became aware of a fact I had not considered before. My mom was a pretty good weather observer. As for me? Well, that is how I began my thirty-seven-year career with the National Weather Service.

In the U S Air Force, I spent four years working as a weather observer and, upon leaving the military, I enrolled in college and spend the next six-and-a-half years working full time and attending classes at the University of Texas at Arlington. State College, a two-year college under the Texas A & M system, and later a four-year university under the University of Texas System. Boy, do I remember those days, nightmares all. I took all the cruddy shifts to go to school.

My Meteorologist in Charge said, "You can go to school, but you have to trade shifts with your co-workers."

I did. I got mostly midnight to eight shifts. I would get off at eight o'clock, sit in class until noon, go home and sleep, before going to work at midnight again. The real nightmares came when I got into advanced mathematics, physics, and chemistry. I still groan when I think of those days. But God was good. Somehow, I got through with my sanity intact. It was my study of fluid mechanics, meteorology, and dynamic meteorology that made the pain worth it. I truly loved learning the intricacies of atmospheric physics. I was intrigued to learn that, if we could see air, we would see mountains, valleys, and rivers flowing, and how warm regions would have higher altitudes for a given pressure than colder areas, since cold air weighs more than warm air. I told you, I gave my memory free rein to go wherever, but—and finally—my mom is part of its ramblings.

I remember a song by Waylon Jennings, "Did ole Hank really do it thisaway," and I say to myself, "Did Momma really do it thisaway?" No, she did not, but she would have been proud of weather knowledge I had acquired in a different way. She had a "sixth sense" about the weather, and she would stand on the front porch and listen for an all too familiar sound which she had experienced as a child when she and members of her family stood pushing on the door to keep it from blowing in during a storm. The sound, she said, was distinctive. To me, it was just plain scary to even hear her speak of it.

Often, she would round up six sisters and me, and order us out of the house and into an old earther cellar where she kept canned fruit and vegetables. When we were all in, my dad would close the door. The question was, which would be worse; to be blown away by a storm or die in the root cellar from snake, centipede, or tarantula bites? The point is this; at six or seven years of age, I would panic every time booming thunderstorms and vivid lightning interrupted our night. We had no "National Weather Service." We had no warnings on radio and we certainly had no television weather personalities because we had no television. Momma kept a kerosene lamp down there and blankets for us kids to lie on. After the storms passed, we would go back into the house having lived through another dreaded storm event.

I wish my mom, the weather observer, had known some of the theory behind what she knew instinctively. But she did not. Nor did I, at that time. So, Momma would issue the warning, the thunderstorms would come, and I would cry. Who would have thought that I would make my living issuing warnings for severe thunderstorms and tornadoes for thirty-seven years? Yep, mothers, thunderstorms, and crying little boys will always be a fond memory for this retired meteorologist.

Depravity of Man

"Oh, the times are changing so fast. I just do not know how much longer the Lord can tarry. The world has never been as evil as now." Yes, it has. Read about it in the book of Genesis in the Old Testament. God created Adam to be spiritually pure. Then, from the rib of Adam, He made Eve also spiritually pure. He set them in a beautiful garden and gave them everything in the garden except the fruit from a tree that He had set in the middle of the garden. He told them if they ate from that, they would die.

Then along came the wily old serpent, the devil Satan, who enticed Eve to eat the fruit from the forbidden tree and she then persuaded Adam to eat it. Out of that one disobedient act, sin was born. Man had fallen, and God placed a curse on mankind. From that day on, every baby born came into the world with what theologians call imputed sin. The condition of imputed sin is kind of like genes that are transmitted to offspring. From that day forth, as a baby grows, its propensity to sin becomes stronger. Left unchecked, man's fallen nature will lead him to commit the most atrocious acts.

There will be no end to the totally depraved things he will do. Read about it in the account of Sodom and Gomorrah also found in the book of Genesis. Oh sure, many individuals live out their life as "good, solid citizens," but they do so not because their nature was good, but because laws restrained them. They obeyed the law as best they could. But, they are still hopelessly lost. The bible says, "all have sinned." Romans 3:23. Here's the dreadful part. "...the wages of sin is death..." Romans 6:23. Ah, but the almighty and loving God did not leave us without a remedy for our sin. "For God so loved the world that He gave His only begotten Son that whosoever believes in Him shall not perish but will have everlasting life." John 3:16.

We now have hope because God put our sin on Jesus and punished Him as our substitute. Now, "...by grace are you saved through faith and this not of yourselves, it is the gift of God—not of works so that none can boast." Ephesians 2:8-9. So, a simple child-like faith in Jesus will obtain for us eternal life. We do not have to earn it by striving to meet all the requirements of the law. We could not if we tried. For any who trust in Jesus Christ, God redeems them from their fallen nature and gives them a new nature. "...if any man is in Christ he is a new creature..." 2 Corinthians 5:17. So you see, even for those who have not put their faith in Jesus, most are constrained to obey the law. They are productive citizens even though sin will condemn their unredeemed soul to hell. "The soul that sins it shall surely die." Ezekiel 18:20.

For my friends and loved ones who have not trusted in Jesus, my prayer is that they will turn to Him and be saved because the insidious nature of sin takes one farther and farther from God. Remember, I said that most folks try to obey the law and are restrained from depraved acts by it? Well, what if there is enough clamor to cause politicians to begin a process of dumbing down laws or even rescinding them. What then? I will tell you. Without laws, the unredeemed soul will sink deeper and deeper into the pit of immorality. Who knows how deep they might sink?

At the Altar of Mammon

I want to begin by stating that God has established the only truth there is. Today, truth has become so warped and unrecognizable. We live in a time when it is so easy to tell "a little white lie." After all, everyone does it. The number of ways that one can "benefit" from adopting a different definition of truth than the one God gave is endless. The unredeemed mind will allow its conscience to be shaped by experiences that have led to "benefits" having been derived from "stretching the truth." Even believers, from old habits, may see opportunities for great wealth if only they will make "this one-time exception" and tell a little white lie or believe one despite knowing that the only truth is founded upon God's Word, the Holy Scriptures. Believers, in their desire for riches often compromise their knowledge of truth until lies seem like truth. There is no truth but God's truth, "Let God be true and every man a liar." Romans 3:4.

Take the matter of desiring wealth. Satan understands that fallen man, even after having been redeemed by the saving grace of God through faith in Jesus Christ, will tend to sometimes have inordinate desires to gain wealth. The only defense against Satan's onslaughts to get us to make wealth a priority in life is by renewing our minds daily by delving into the scriptures. "…And do not be conformed to this world, but be transformed by the renewing of your mind, so that you may prove what the will of God is, that which is good and acceptable and perfect." Romans 12:2 A diligent search of scriptures will always point in the direction of truth and thereby direct a believer away from satanic lies.

Satan is wily. He will tell you the voice you hear that says "go for the gusto" is from God and, without a solid foundation in the true word of God, you will fall for it.

"But," you say, "I know God is in this plan of mine to be rich to help the poor."

Is He really? Why, then, did He tell the disciple who complained (probably Judas), when the woman poured a vial of expensive perfume on His head, "…the poor you have with you always, but you do not always have me"? Matthew 26:11.

So, here is a true scriptural test of the soundness of your seeking for wealth. If you spend more time planning and scheming how you can get rich than you do studying the scriptures, then you are not serving God and God's voice is not the one you are hearing. "No one can serve two masters. Either you will hate the one and love the other, or you will be devoted to the one and despise the other. You can't serve both God and mammon (money)." Luke 16:13.

Are you struggling to acquire riches, yet making no headway? How long has it been since you spent time in God's word and in prayerful communication? Do you say, "Lord thank you for the food you provided, Amen"? Be careful that you are not seen serving Satan because of your love of money. Money is not the root of evil but the LOVE of it is. 1 Timothy 6:10. Turn to the truth of God's word so the Holy Spirit can feed your soul and in turn give you discernment. Stop worshiping at the altar of mammon. God will provide for all your needs.

"Look at the birds of the air; they do not sow or reap or store away in barns, and yet, your Heavenly Father feeds them. Are you not much more valuable than they?" Matthew 6:26

They Died So Young

So very many young soldiers who fell on the beaches of Normandy, France never saw another Thursday after June the eighth, nineteen-forty-four. They gave their life for our freedom.

We had just come from viewing the memorial at Utah Beach. The wind across the English Channel blew so strong it was impossible to keep a hat on. The sand was biting in its fierceness. We had to pull our jacket hoods up over our heads and walk backwards going there.

A few minutes later, I was remembering the sting of the sand on my skin and felt ashamed that I even took note of it. As I stood at the site of the Normandy American Cemetery and Memorial and viewed the more than nine-thousand white crosses where young men and a handful of women were laid to rest, almost overwhelming sorrow struck me. Most were eighteen and nineteen-years-old.

When General William Tecumseh Sherman said, "War is hell," he understated it. I could almost hear the cries and screams of our young men as they fell like flies under heavy machine gun fire, possibly originating from the very spot where I stood overlooking Omaha beach and the English Channel where most of them died.

I thought about what might have occupied their minds as they sat silently on the troop boats waiting to land on the beach. They say the beaches were quiet until the first landing craft ramp was dropped and then all hell broke loose.

I remember thinking, "I hope they had viewed this as an exciting adventure while on the boat, and then died before they realized it wasn't." I thought I had been touched as deeply as one can be by the war memorials in Washington D.C., but—there—I did not stand in the very spot where it happened.

Oh Lord God. Please keep America free so that this never comes to our beaches. God bless our veterans and fighting men and women.

Coffee, The Right Way

I just received a new order of green coffee beans from South America delivered to my door. Just now, I transferred fresh roasted Colombian Huila Valencia beans from my home roaster to my coffee grinder, then to my "Moccamaster" coffee maker that has a temperature setting of two-hundred-and-five degrees Fahrenheit which is the optimum setting for excellent coffee. I used filtered water, six-ounces per cup and two tablespoons of ground coffee per cup. Myrtle, it just doesn't get any better than this! Wait! Who's Myrtle?

From Hangnails to Hangings

When we lived in Anchorage years ago, we had a friend (church member) who was famous for saying, as his mother taught him, "God cares about everything from hangnails to hangings." I have grown to realize the truth in his mom's observation.

There was once a time in a less mature stage of my Christian walk that I would never have considered that to be the case. I did not want to ask God for trivial things. I considered asking for small things to be an imposition upon the Almighty's time.

I can't affix a time when my views about that changed but they did. Now, sometimes, when I am about to be late for an appointment and can't find a parking place, I have no hesitation to ask, from my heart and not just vocalize it, "Lord, I praise you for caring about the small things in my life. Please cause a parking space to open for me." I can't count the times, after driving around the block and sometimes even in the next few yards, a car parked in front and to the right of me begins backing out.

Oh yes, my Heavenly Father cares about the wee tiny things in my life because He loves and cares for me. Just as I was with my kids, only on a vastly incomparable scale, God, my father, delights in giving me the things that are good. We know that, "every good and perfect gift is from above and comes down to us from the Father of heavenly lights who does not change like shifting shadows." James 1:17. And in the larger things of life, He operates the same way because my God is not wishy washy. He is steadfast and faithful.

"But I do not want to pester God," you say.

Well, He does not get frustrated or irritated. Remember the parable Jesus gave about the importune widow found in Luke 18:1-8? A widow went to a judge to get justice from her adversary, but he was a man who did not fear God nor regard man, so he did not grant her petitions. She kept going back again and again until the judge gave in and granted what she asked for. Jesus' point in that parable? Do not stop praying for something that is good and right even though it might seem your prayers go unanswered.

As we strive to revere God more day by day, and learn more about Him, the clearer it becomes to us that God's timeframe and methods are not easy to understand sometimes, but they are always perfect. He does not delay because He is trying to figure out what is best for us. He does not delay because He is frustrated at our asking. No, God is not like us. Nothing rattles Him. Nothing worries Him. Nothing causes Him to doubt Himself. He is just being God, our sovereign God. "For my thoughts are not your thoughts, neither are your ways my ways, declares the Lord." Isaiah 55:8.

So then, the next time you feel that your problem is too small for God to care or you feel you are imposing upon His time and He is busy, remember God is eternal and timeless. Our greatest problem is no more difficult for Him than our smallest. Difficulty is not a word that applies to God. Go ahead and ask Him to help you find your lost car keys. After all, God cares about everything from hangnails to hangings.

Big John and Me

After driving for what seemed like hours along very narrow roads in the Ireland countryside, we finally reached the object of our search, a bridge.

Ah, but it was no ordinary bridge. This bridge was special for it was where John Wayne and Maureen O'Hara filmed "The Quiet Man."

I sat down on the bannister of the bridge, in the very spot where John Wayne sat, and—for a fleeting moment—I was the Duke.

The Greatest of Conflicts

Oh, my dear friends! I hope that in your lifetime you never have to struggle and be conflicted emotionally for any reason. It is hard to live in today's world without—in some form—being distraught over what choices we should make.

This morning and all day long, I have been burdened because of actions that I took. I struggled with what choice to make. And this afternoon, after raking my leaves in the yard, I struggled with guilt.

You see, I was conflicted this morning because I had planned to rake the leaves in my yard, but I heard on the news that environmentalists are telling us we should not rake our leaves. Why should we not rake our leaves? According to those environmentalists, when we rake our leaves, we destroy thousands, perhaps thousands upon thousands, of tiny creatures who make their home beneath the leaves in our yard.

Here are just a few of the valuable things that live so comfortably in the houses we provide by not raking the leaves: cinch bugs, fire ants, army worms (you know the kind that get in your tree and make webs that drop grubs on your patio and kill the leaves on your tree), Bill bugs, chiggers, and cutworms just to name a few. How is it, that I could be so cruel?

The environmentalists would plead with us that the little critters who live under the leaves in your yard are far more important than the house you live in, the house that might burn down if all those leaves catch fire. Well, I have agonized all I am going to about this grave dilemma of whether to let my house burn down or clean up my yard and destroy thousands of homes for critters. I opted to rake the leaves, and bag them, and set them on the curb.

I think that, perhaps, I shall lose this heavy burden of guilt in a few days if one of the little critters does not turn out to be a snail darter. I do not think I could bear such guilt. Oh wait! Snail Darters live in water, don't they? Oh, how good I feel now! I did not kill a snail darter.

Of Fireflies and Fruit Jar Lanterns

In the book, The War, An Intimate History ;1941-1945, by Geoffrey C Ward and Ken Burns, Katherine Phillips gives an account of life in Mobile, Alabama during World War II.

She states, "The pace of life was slow. There was no air conditioning, of course, so on a hot summer evening Daddy would load us in the car and we would drive downtown to Brown's Ice Cream and he would buy us an ice cream and then we would drive out to Arlington and park by the bay. We would sit there and enjoy the sea breeze. Then when we had cooled down enough he would bring us home and everybody could go to bed and sleep. Or, we would sit on the porch in the evening and the children played in the yard. It was a wonderful way to grow up. Down in Mobil we were completely away from the rest of the world."

In Panola, Oklahoma ours was a similar experience. Adults sat on the porch in the evening, fanning with a piece of cardboard, talking about things of interest until the temperature lowered enough to go inside. It was a way of life back then, in southeastern Oklahoma, in the summertime. We children loved it. We ran around in the yard catching lampyrids (biological name for lightning bugs). There was always an abundance of the little flying creatures. We were so cruel (without knowing it). We would smash the glowing little bug against our clothes and rake downward, leaving a bright neon light streak on our clothes. Then, having labeled ourselves as aces among firefly hunters, we proceeded to imprison our catch in fruit jars. We enjoyed the lanterns we made.

Often, we would bring our "lanterns" and set them on the porch railing and then seat ourselves near the adults either on a banister or on the floor.

Our proximity to them was dependent upon the conversation in which they were engaged. Sometimes they told ghost stories, for our benefit I think, and we would scoot closer to our parents. Of course, they would not let harm come to us. So, we continued living the good life unaware that far, far away over the ocean, our uncles and brothers were fighting to keep America free. We could go inside and sleep on comfortable mattresses, albeit sometimes we found it necessary to sprinkle the sheets with water to get some relief from the heat. They, on the other hand, slept in foxholes, often muddy ones. I doubt they slept much though.

I am eternally grateful to the men and women of all wars, heroes all, because they kept evil warriors from coming to our shores. Nearly two-thousand died on the beaches of Omaha and Utah in southern France during WWII, the war of my brother and brothers-in-law. Their blood soaked into the sands so that we children could enjoy our fireflies and fruit jar lanterns.

Who Is Righteous?

" There is none righteous, no not one. There is none who understands. There is none who seeks after
● ● ● God." Romans 3:10,11.

In this part of the apostle Paul's letter to the Roman church, he refutes the Jewish belief that they had a special relationship with God that would make them not guilty before God based on their works and what they believed to be a greater standing than the Gentiles had. He was showing that all have sinned. All needed God's saving grace. My eyes and my heart keyed on "...There is none who seeks after God," in verse 11. A sadness flooded over me as I thought about my sin, both past and present.

How great God is that he loved me during my sin and extended saving grace to me! How great He is that, even now as one redeemed when I do sin, he forgives me! His forgiveness is limitless. Paul told the Jews that, just like the Gentiles, they were sinners and needed forgiveness. He said where sin is, then grace abounds to the glory of God. But, anticipating that the Jews would see his statement as an occasion to sin more, he told them if they did sin for that reason, they would deserve the punishment they received. His purpose was to show the opposite of what they believed. Paul's intent was to show that glory accrues to God when we sin because it reveals His power to forgive sin and to change the believer into a new creature who is seen by the Father to have the righteousness of Christ.

This, of course was God's plan since before the world was created. We inherited our sin condition from Adam, who sinned in the garden of Eden and was cursed by God so that everyone born into the world is condemned to eternal punishment. Further, we can do absolutely nothing in the way of working to cause God to accept us into heaven. "The wages of sin is death but the gift of God is eternal life through Jesus Christ the Lord." Romans 6:23.

It has been said that man seeking God is at the core of every other religion in the world, except Christianity where, in truth, God seeks out man. How desperate we are without even knowing it! We are born into this world and immediately set upon a road that leads only to hell. That is, until we hear the good news about Jesus and learn that He is "...the way, the truth, and the light and no man come to the Father except through Him." John 14:6

When Christ died on the cross, he took on our sin and God, the Father, punished him in our place. Because of His sacrifice, we are made right with the Father. And it is all free. "For by grace (undeserved merit) you are saved through faith (in Jesus) and that not of yourselves: it is the gift of God, not of works lest any man should boast."

So, why does man not seek God when they are in such great need of a relationship with Him? In many cases, it is because they have not been told about Jesus and the fact that he saves us from our sins. "How then can they call upon Him whom they have not believed? And how will they believe in Him whom they have not heard? And how shall they hear without a preacher?" Romans 10:14.

Those of us who have placed our trust in Jesus, and have secured for ourselves a blissful eternity in Heaven, have a command from Jesus, "Go into all the world and preach the gospel (good news)." Mark 16:15.

For some, "all the world" means just that. Go as missionaries. For most of us, "all the world" can be one individual at a time. Whatever "all they world" means, we must obey. How can they believe without a preacher?

Sound the Alarm

We are all going to die! We are all going to die! The sky is falling! The sky is falling! We are all going to die, but probably not before we all must walk to every destination that we choose. Why? Well the climate change crazies are clamoring for EPA regulations far beyond where they go now. Friends of the Earth and Sierra Club, for example, do not want us to fly in airplanes or drive cars because the emissions damage our climate.

Now isn't that just special? If we walk to our destinations, I certainly hope that nobody is monitoring all that nasty old carbon dioxide that we exhale. It is just horrible. We could be in serious trouble.

Oh, but of course there will be exceptions made in the regulations. The bigwigs will still fly in their private jets while we peons will walk. They are entitled, don't you see? Makes me want to puke.

Even had I not come out of an atmospheric science background, I would have no trouble in evaluating their efforts as pure unadulterated hogwash. In my opinion, not one of these morons have the vaguest clue what is involved in the destruction of our planet. God says he will destroy the earth by fire and brimstone. Oh, well now, perhaps that is what they mean by global warming. Sound the alarm.

For the Unborn

"I will praise you because I was fearfully and wonderfully made. Thy works are wonderful. I know that full well." Psalm 139:14.

Why were you and I born? Was it because God was lonely? No. "The God who made the world and everything in it is not served by human hands, as if he needed anything." Acts 17:24.

Why then?

The reasons are too numerous to list and reference here, but they all boil down to this: He wanted us to worship and glorify His name in all the Earth. Praise and honor can't be given to God by the dead; neither the physically dead, nor the spiritually dead.

A couple of days ago, I posted about Planned Parenthood having been caught red-handed discussing prices of aborted baby parts. I used the term fetus. I herewith repent of having used that word. "Fetus" is a substitute word. It salves the conscience of baby killers. It seems that it is easier for them to say, "If you don't want children, then we recommend that you have an abortion. After all, it's just a fetus."

No. It is not just a fetus. It is much more. It is a living human being and abortion is not just getting rid of a troublesome fetus, but rather it is the determined act of killing a baby—a human being. Interesting, isn't it?

In Texas and many other states, harming a child draws severe punishment. Kill a baby and you are sure to "get the needle." But if you can use forceps and crush the baby's head before it leaves the womb, then it is abortion. You just got rid of an unwanted fetus.

I wonder if the doctors and nurses ask themselves when aborting a baby, "Would this baby have grown up to become a servant of God? Would he or she have praised God and brought glory to His Holy name?" Probably not. We take care of infants in our church nursery, and I thank God for each little one I have the privilege of holding, for still another voice to praise God.

Oh, The Lion Roars Tonight

"Be of sober spirit, be on the alert. Your adversary the devil prowls around like a roaring lion, seeking someone to devour." 1 Peter 5:8

Did you ever notice how non-believers in Jesus Christ seem to go through life without having many hardships? Of course, this is a false perception, no doubt, but nevertheless it often prompts believers to say things like, "Why do I struggle so, when people curse God and seem to enjoy life and all it has to offer?" It's true. They do. However, believers will find their rewards in heaven throughout eternity, but—in weak moments—we still fear and are anxious about things.

When I was a child I had no worries, nor concerns. I am sure my parents must have had them, but we younger children—as far as I remember—never heard them speak of the matter. We ate abundantly from a huge garden. Momma was a great seamstress and fashioned our shirts and the girls dresses out of flour sacks (they sold flour in some mighty pretty sacks back then). We went bare-footed as early in the year as we could and continued to do so if weather permitted. What else was there to worry about?

Today, unlike the yesteryear of our youth, there are so many things we could, and do, worry about despite Jesus' admonition, "Be anxious for nothing." Yep, me too. My most recent bout with worry occurred about six months ago, when someone hacked into my PayPal account and ran up over four-thousand dollars of debt. I did not incur the debt and I knew it, but PayPal insisted that I did. It took several months and many sleepless nights before it got resolved, thanks largely to a lawyer friend. Should I have worried? No. But pride brought with it anger at having had such an injustice perpetrated upon me.

The sudden news of stage four cancer—which one of our friends is now facing, news of burglaries—such as one friend recently experienced, assaults, financial stresses, stock market crashes, attended by financial losses, and many other things sometimes cause us to almost despair. Do not. Can you change anything at all about circumstances that are out of your control? You can't. Jesus said that in the book of Matthew during His Sermon on the Mount. Matthew 6:25-29. What are we to do then in such circumstances? We must trust the One who can fix such things, our Lord Jesus Christ. His power is boundless. Fear and doubts shout that our faith is weak, and we are just not quite sure God is up to the job of effecting a fix.

Something that helps me in such times is to recite, to myself in the night, the Lord's Prayer. Our Father who art in heaven. Our Father sits on His thrown with an overview of the universe. He never sleeps, nor slumbers. Psalm 121.4. As parents, we looked in on our children in the night and went to them when they cried. Is our heavenly Father less a parent than us? "…and thy thoughts which are to us-ward; they can't be reckoned up in order unto to thee; if I would declare and speak of them, they are more than can be numbered." Believer, do not be afraid in the night. God is watching over you.

And Jesus' prayer goes on to ask the Father, "Give us this day our daily bread." Remember the daisies? If He provided for them, then will He not also provide all your needs? So many scriptures in the bible promise that He will indeed. Oh, how I wish you did not have to go through what you are now enduring! But, in every trial of life there is a lesson to be learned. Ask God to show you the reason for your suffering. Until you get an answer, trust Him. Love Him with all your heart, soul, and mind.

He will see you through this. "Cast all your anxiety on Him for He cares for you." 1 Peter 5:7

Remembering Snip

He stood fifteen-hands high, this majestic horse of mine. His eyes were ablaze when he sensed that he was about to carry me over the Oklahoma hills and across the plains. Countless days were spent roping strays and fighting off rustlers. Roy Rogers could not have done it with more purpose, and—I might add—charm. I was a hero to all the ranchers in the Cattleman's Association. But I could not have done it without Snip. He had an uncanny knack for spotting trouble.

Once when we were out looking for strays, we came upon three men skinning a steer. Snip could not talk, but if he could have, I am sure you would have heard him say, "Shazam," just before tearing into that small group of scallywag outlaws with a look of the devil himself.

I can still feel his long mane brushing my face as, together, we raced toward those owl hoots with his front legs slapping forward to pound them into dust.

But wait! Childhood memories got the best of me. Snip was not anything special. He was just Snip, an ole plow horse. He would not have placed tenth in a field of five show horses, but he did give me many joyful days as I sat astride him wearing my dad's big rubber boots that came to my knees. I was so skinny that I needed the extra weight of those boots to keep me from falling off when he galloped.

Snip was a good horse, but oh how frustrated he could make me. He did not like to be caught. I must have chased that horse a million miles before I discovered that a bucket of oats would do the trick, dumb kid that I was.

To put a bridle on this old sorrel, I had to climb up on a wood slatted fence. Once bridled, I would throw my leg over his back and off we would go, Snip and me. No, he was no "drinker of the wind," but he was my horse.

For the longest time, I rode him bareback.

Once, we were in Fort Smith shopping, a real treat for my mom because her usual shopping consisted of going to Wilburton occasionally. Compared to Wilburton, Fort Smith was New York. While there, I spent a long time outside a tack store looking at saddles. I guess my dad noticed because some time later a friend offered to sell me his beat-up old saddle for five dollars. I asked dad to buy it for me. I can still hear him pose his question to me.

"Do you want me to buy it for you now or would you rather wait until I get the money to buy you a brand new one?"

I said, "I'll be happy with the five-dollar one."

And I was.

See? I told you I was a dumb kid.

I smile when I think of that day. I bet my dad breathed a sigh of relief. Ah, but my relationship with Snip came to a tragic end. One day I came home from school and dad told me that Snip had gotten out and was hit by a train. He was dead.

If there are horses in heaven, then I know Snip is there. Perhaps he will return one day with Christ, who will be riding a white horse according to the book of Revelations. Snip will be proudly riding by his side into battle. Until then, Snip, my friend, drink the wind.

Growing Up Okie

I was listening to a Charlie Pride song this morning. You rednecks know it. It's called "Cotton-picking Mississippi Delta Town." Charlie kind of sung my life, except I did not grow up in Mississippi.

Eastern Oklahoma in the nineteen-forties and fifties (my growing up years), was like many southern states. Some things about the culture back then are best not remembered, but—for the most part—growing up poor was a blessing. I never knew we were poor. My dad always managed to keep a roof over our head and food on the table (well, Momma put the food on the table for a family of eleven; she was a great cook). Over the years, as my faith increased, I have come to realize that my parents never put food on our table. Nor did they keep a roof over our head. God, alone, did that. He just let Momma and Dad take credit.

We had three milk cows. Momma milked two, I milked one. We had lots of eggs and butter, and on Sunday we often had fried chicken, especially if the preacher was coming to our house after the service. I remember Momma always chose the back. She said she liked the back, but I was not fooled. There were only two little morsels of meat on the back. I knew she was putting her family first. It was her nature.

Dad always took milk and butter to a family in the community. I sometimes wondered why he did, because the man—a Socialist—once told Dad that our cows were just as much his as ours. In retrospect, had I not been a kid who was taught not to be disrespectful, I would have said, "Well, Mr. (name redacted), why don't you come around at milking time and help with the milking chore?"

Momma made beautiful quilts and occasionally she would invite ladies in the community over for a quilting party. She would let the quilting frames down from hooks attached to the ceiling and the women would quilt and chat. I guess the living room could have been referred to as a "chat room."

Dad raised cotton for a cash crop and Momma and some of my sisters would "card" the cotton. In other words, they would put cotton balls between two wire brushes and rub back and forth until a nice sheet of cotton was ready to be used as batting for the quilts. Some of those quilts were heavy and came in handy on those cold Oklahoma winter nights. Momma collected the feathers when she killed chickens. We slept on feather beds in the winter.

We lived in a house at the foot of a ridge. On the other side, of which was a creek called Big Fourche Maline Creek, Dad and I would often set out trotlines to catch some big channel catfish. The next day we would climb the ridge and walk down the other side to see what we caught. I still like catfish to this day.

Dad always raised a huge garden and Momma canned fruits and vegetables in a large pressure cooker. She scrubbed our clothes on a rub board, under giant cottonwood trees down by the spring. She would build a fire under a large black kettle and heat water drawn from the spring. Sometimes she would dig sassafras roots in the springtime and make tea which was believed to be good for one's health. At the end of the month, dad would pay the grocer who would put enough bubble gum in a sack for us kids.

No siree Bob, we were not poor. We were rich beyond expectations. Hey! When I was in high school, during summers, I made five dollars a day working for the county. I helped repair bridges and handed out commodities (butter and cheese). Ah, the good ole days!!

Do I want to go back to those days of reading by kerosene lights? Uh, do I have to give up my computer and Facebook?

Of Drylines and Texas Thunderstorms

Back in the nineteen-fifties, few meteorologists had ever heard the term dryline, dew point front, or Marfa front. These are all a nomenclature most common to Texas, Oklahoma, and Kansas. As our population grew, and more and more television stations began broadcasting weather reports, knowledge of the phenomenon also grew. Although the term "dryline" is now heard around the kitchen table especially in spring, few people (apart from students of the weather) understand the complexities of one of the most frequent drivers of thunderstorms in this part of the country.

Meteorologists analyzing weather maps will locate the "dryline" or "dew point front" at the point where moist air suddenly becomes very dry at the surface of the earth. This is usually near the fifty-five degrees Isodrosotherm or line of equal dew point temperature. It is not uncommon to see dew point temperatures drop from fifty-five degrees to the twenties and teens, and occasionally to the single digits as air moves down the eastern slopes of the Rockies, warms up, and dries out from being compressed due to higher pressures.

Because dry air is heavier or more dense than moist air, the dry air pushes eastward under the moist air, thereby lifting it to the convective condensation level where it initiates thunderstorms. Yes, I know it sounds wrong, but trust me. It has to do with the fact that dry air contains more nitrogen than moist air and nitrogen is heavier than oxygen. It makes sense, doesn't it? The two main gases in the atmosphere are about seventy-eight percent nitrogen and twenty one percent oxygen.

Unlike ordinary fronts moving in from the west, dry lines will not continue to move east since they are sun driven. Elevations increase to our west forming an inclined plane. Obviously, the deepest Gulf moisture will be found in our area and shallower out west. As the sun begins to "burn off" the moisture to the west, the dry line will advance eastward. At night, lacking sunshine, it will advance back to the west. Once a dryline sets up out west, this back and forth movement will occur day after day until an upper air feature links up with it and carries it on to the east.

So then, the next time someone drops the word "dryline," just say, "It is not that simple, Myrtle."

Praise Jesus

Oh, the wondrous love of Jesus, that He would die for me! For me to earn my way to heaven, I would have to live a perfect life. I could not. He did. Then, on the cross, it pleased the Father to credit my sin to Him and then punish Him in my place. All I must do is believe on Jesus' name and all my sins, past, present, and future are forgiven. Praise the wonderful name of Jesus.

Another Visit to the U.S. Virgin Islands

As I stood on the dock at Charlotte Amalie, St. Thomas waiting for the ferry to St. John, a flood of memories rolled over me like swells coming ashore. It was the same feeling I got each time we returned with my family.

My first visit to the beautiful U.S. Virgin Islands was in nineteen-seventy-one when I went there as a Hurricane Preparedness Meteorologist for the Caribbean. I was stationed in San Juan, Puerto Rico. Our children were small. Carrie was only eleven months old. We have since returned to the island numerous times with our now grown children, and their children. As we climbed on board the ferry, I breathed deep to get what might be my last smell of salt water. A week at Villa Panache, high in the mountains of St. John, afforded me time to soak up the sights and sounds that inhabit that majestic, yet simple, island. At eighty years of age, would it be the last such trip for Rheta and me?

As a meteorologist, I did a survey of exposed populated coastal areas throughout Puerto Rico and the U.S. Virgin Islands. Back then, National Weather Service personnel and Federal Aviation Administration employees could fly round-trip on a "space available" basis to the U.S. Virgin Islands for five dollars. Rheta and I both smoked in the nineteen-seventies and she would fly to St. Croix and, sometimes, St. Thomas to buy cigarettes and jewelry tax free. We both quit our smoking habits long ago.

I was required to take a crash course in Spanish, so Rheta took the class with me. We both made some funny mistakes as we practiced our Spanish. One day, Rheta and her friend, who was a native of Puerto Rico, were shopping in a jewelry store around noon. Rheta wanted to practice her Spanish so, since it was lunch time, she said in front of the store manager, "*Yo tengo hombre.*"

Curious as to why the store manager snickered, Rheta asked her friend, "Did I say something wrong?"

"Yes," her friend replied. "You said that you have a man. You should have said, '*Yo tengo hambre.*'"

Rheta tugged at Maria's arm and both made a hasty retreat.

Of Screen Doors, Latches, And Prankster Sisters

"Some trust in chariots and some in horses, but we trust in the name of the LORD our God." Psalm 20:7

Funny how the older one gets the farther back in the recesses of the mind one must go to revisit childhood memories. Ah, but so very worth the effort is that mind journey. One of the things I often think about is the pranks my sisters used to play on us kids and themselves. I was blessed to be the last of nine children, seven of which were girls. I admit to having been spoiled, but I also insist that the pranks they pulled were not always fun. In fact, as I think about it, we were all pranksters, except Momma.

My dad smoked a pipe and, when I was very small, he would send me to the kitchen to get him some matches. When I came back, he would be on his hands and knees growling like a rabid dog. He had thick curly hair and I made good use of it. I would grab both hands full of his hair and pull. I am sure the pain I inflicted upon him saved my life many times.

My sister, Nadine, once put a stocking over her head and knocked on the front door. She looked hideous. My dad opened the door and momentarily looked startled, but then he reckoned that this fellow, whomever he was, needed help.

"Well, come on in," he invited.

We all had a laugh. I am sure we all got our talent for playing pranks from him, but my sisters did it best.

As my bedtime drew near, one or the other of them would crawl under my bed and wait for me to come to bed. Then, the most demonic sounds would issue forth from the troll-infested den under my bed. I would pull the covers up over my head and there find refuge.

We would all go to bed at night and one of the last things Momma would do was to make sure all the screen doors were latched. Summers were hot, and the doors and windows were always left open. Only a small metal latch kept the screen door from being opened. I never felt safe if the screen door was unlatched. Funny what we trusted in to keep us safe back then.

Well, I doubt that very much trust was placed in the door latch by anyone but me. I suppose the point of all this is that trusting in anything except the name of the LORD, our God, is more futile than latching a screen door or pulling covers up over our face expecting to be safe.

"Some trust in chariots and some in horses, but I shall trust in the name of the LORD our God." Psalm 20:7

Do Not Get Cozy with a Macaw Parrot

His name was Blue, and I loved that bird. He was so funny. Some years ago, Rheta and I got interested in parrots and all kinds of other birds. We did some bird watching whenever we got a chance. The more bird shows we attended, the more I wanted to buy every bird I saw, particularly the parrots. We ended up owning several, big and small. At one time, we had two Macaws, one Eclectus, one Jenday Connure, a cockatiel named George, and Red Factor Roller Canary.

They were wonderful entertainment. We enjoyed them for several years. Then reality hit us smack dab between the eyes. These things were lots and lots of trouble, and they are going to live to be seventy or more years old. Keep that in mind if you ever want to acquire one. If you do ever get one, make sure you do not put its cage too close to the wall. If you do, you will come home and find your wall in need of repair. I thought horses were bad about cribbing after having a barn chewed almost to the ground by my Arabian horses. But, I think these danged birds could out-crib any horse.

All the birds were beautiful and interesting. One of the big Macaws was brilliant red and gold, but he was mean looking and a bit temperamental. Because of that, he did not get as much of my attention as did Blue, the big blue and gold Macaw. He was special, and he liked me. Well, at least I thought he did until, one day, I took him out of the cage and proceeded to continue his training program I had started with him. I was going to teach him to kiss me on the cheek like our Jenday parrot, Cricket, would do on command.

On this day, I guess Blue was just not in a kissing mood. I took him from the cage, holding him by both feet. He was a large bird and had a pretty good reach. I pulled him forward and issued the command, "Blue, give me a kiss."

At that point, he leaned forward as though to kiss me on the cheek and then promptly bit me all the way through my right nipple. It was the last time I ever tried to teach Blue tricks. I promptly called Rheta and asked her to stop at a jewelry store on her way home and buy me a nipple ring. I thought I might as well get some benefit from my newly pierced anatomy.

Blue would go for days on end without talking, except that he never failed to greet me in the morning with, "Hi Blue," which—of course—is how I greeted him, and, "Goodnight Blue," when I turned out the light at night. He was a real corker, and I think he knew it because he sure kept us entertained.

We had a Doberman named Liebchen, whom I would let out the back door to do her business. The door opened to the back yard from the same room where we kept the birds. Liebchen loved the back yard and every time I let her out she would lunge through the door, practically knocking me down. Every time she did that I gave out with, "Stupid dog."

This went on for days until my patience wore thin. I hastily devised a plan. The next time I would pretend to open the door and then shut it as she lurched forward. So, when it was time to let Liebchen outside again, I put my plan into action. I pulled the door open just enough for her to anticipate the open door. As she sprang forward, I shut it and she went headlong into the door. Before I could say it, Blue said it for me, "Stupid dog!"

A couple of years have passed since we found good homes for our birds and now I am well past missing them. Oh sure, we think about them when we sit on our patio and watch smaller birds feeding from the bird feeders and drinking water from the birdbath. I am glad we had the experience. Oh, and the nipple ring—well—I did not really get one; although the part about my first and last piercing is true, I just embellished the story with the part about the ring.

Eventually Time Leaves Everyone Behind

What do the names William Oughtred and John Napier mean to you? Well, I doubt they mean anything to you unless you majored in mathematics or one of the sciences long years ago.

The other morning, I needed to make some calculations as I was setting up an Excel spreadsheet to help me keep my sister, Nita's, financial records organized. I reached for my iPhone and accessed one of several calculators on it. Suddenly, my mind drifted back to my college days when I studied mathematics and physics, but it did not stop there. It kept on going until finally it stopped at one of my elementary classes in arithmetic where the subject matter was long division. Oh, how I hated the tediousness of that method of obtaining a quotient!

I must not have been the only one who did not like long division though. As far back as the fifteenth century a man by the name of John Napier discovered something called logarithms. He found out that by adding logarithms one could arrive at the product of two numbers when multiplied together. Of course, the corollary that by subtracting the logarithm of two numbers, one could find the quotient was also true. He invented a rudimentary calculator that became known as Napier's Bones. It would multiply and divide numbers, and later improvements enabled one to take square roots, but it was cumbersome to use.

Then, some years after Napier's contribution, a fellow named William Oughtred and others, building on Napier's work, invented the slide rule. It revolutionized the world of mathematics and science. It was a wooden device made from bamboo. It resembled a ruler, only wider. There were numbers all over it. It had three parts. Two fixed scales on the top and bottom, with a sliding scale in the middle. There was also a sliding piece with a hair-like mark that linked the other numbers together in their proper relationships. By lining up the marks on the sliding part with marks on the top or bottom, one could multiply and divide. It was also possible to square numbers and take the square roots. There were slide rules that enabled one to deal with trigonometric functions, but they were expensive. I had not sold enough eggs and butter (tongue in cheek) to buy one of those. Well now, is not that the luck of a redneck boy whose most advanced math class in high school was taught from a book <u>Higher Arithmetic</u> by Stone and Mallory? Did the book discuss mantissas? I don't recall.

At any rate, when I declared mathematics as my major field of study at the University of Texas at Arlington, I went out and bought myself one of "them there slip sticks," and suddenly I was transformed from a country boy to a highfalutin mathematician. Throughout my first three years of higher education, I used that slide rule and thanked God that I did not have to work out my calculations in long division. So, there I was, straight off the farm with the smell of cow poop still fresh on my shoes. Ah, but it was okay because I was a bona fide, certified, wannabe mathematician.

In the last year of school, Texas Instruments began selling the revolutionary calculator. In the late nineteen-sixties, I paid one-hundred-and-twenty-five dollars for it. Using that handy device, I could add, subtract, multiply, and divide. I could also square a number and extract square roots. Of course, nowadays, I have an app on my smartphone. Now, the sky is the limit. I can literally use the scientific calculator to express, in scientific notation, distances to stars that are light years away. Or I can find the future value of investments using the financial calculator that is resident in that one iPhone app.

I still have my old slip stick. It is tucked away somewhere in the same old brief case I carried to school. One day, perhaps, I will pull it out and show it to a great-grandchild. Although, I suspect they will say, "Thank you, grandpa, but I'll use my calculator." Then they might add, "Grandpa, tell me again. What is a logarithm, and who were William Oughtred and John Napier?"

Not Really

I ran across this poem while cleaning out our garage. I wrote it as we prepared to leave the land of the midnight sun. I post it now for Neil Thompson, who pastored Sunset Hills, and for Geraldine McLaughlin, who lived in Bear Valley. The year was nineteen-eighty-two. We had said goodbye to our Anchorage, Alaska friends, and had worshiped the last Sunday night there at Sunset Hills Baptist Church. We heard Art and Barbara Braendel sing their last duet. We were Texas bound. More than three decades have passed, yet I can still get lost in my memories of the way home from church and, though I have driven it many times in my mind, I really did not. So, I have entitled this "Not Really."

"Not Really"

We turn right out of Diamond Center.
A Tosoro station is on our left.
Klatt Road is next,
Then Huffman Center.
We stop for a bagel
With loks.
Turn right on Oceanview.
Cross the tracks.
Nora is next—twelve eight three one,
But not really.
We peer down into

A white shrouded Bear Valley.
Only days ago
The hills were brushed
With greens and golds.
Now white—deep white
Yet majestic.

The road is icy,
So slick!
We dread going back
Up the hill.
It was fun though,
Being with friends.
We must do it again
Tomorrow,
But not really.

Look!
There's the yellow house.
A nose is pressed
Against the upstairs window.
Familiar.
They call the wind Mariah.
Lasagna!
The fire was warm.
The conversation—
Gentle.
I'm full now.
But not really.

I must take care
When turning left.
Drive on around
To the back.

We're late.
I hear singing.
The sermon was great.
Let's stop awhile,
At the parson's house.
You talk.
We'll watch the game,
But not really.

The airplane is leaving.
Quick! Get aboard.
It's a long drive from
Seattle.
But we made it home.
We left the beauty
Of a majestic land
Behind.
But not really.

Sannado

We sat in the stands at an Arabian horse sale ring in Waco, Texas, Rheta and I. There were thirty or so other horse enthusiasts there, as I recall. Excitement was running high as we all listened to a conference call where several moneyed individuals were bidding on a Bask daughter. You see, Bask is the Arabian horse industry's standard, as many see him. I do not recall the final bid, but it was more than twenty-thousand dollars.

I looked at Rheta and asked, "Why are we here?"

She replied, "I don't know, but let us see it through."

Several offerings were let loose "at liberty" in the ring, and the bids began. None fetched anywhere near what the Bask daughter did. After a few minutes, a beautiful, rose-colored gray, two-year-old colt trotted into the ring. His handler beat on a drum and Sannado strutted his stuff. That fantastic looking youngster skip-danced all around the arena as the crowd shouted their approval. The bidding started. We kept quiet until it became apparent that, although the crowd loved this Bask grandson, the market that day was not boys. Several mares had sold quickly, but only two individuals were bidding on Sannado. We jumped in. A bid was given. We countered. Another bid was made and, again, we countered. No further bid was made. We had bought ourselves a purebred Toronado son and Bask grandson.

Now what? We had a horse, but we came in a car. We did not have to wonder long, however, because the sales company had thought of what we had not. They had a truck and driver who would transport the horse to Springtown where we lived. They would follow us.

We got home around ten PM. The truck could not navigate the ninety-degrees turn into our gate, so we unloaded Sannado and I walked him a couple of hundred yards to the barn. I put him in a pen with a stall, scrubbed his ears and mane, and walked away.

As I opened the gate to leave the pen, he promptly trotted over, bit me where I sit down, and quickly ran back to his stall. The bite was not hard, and I knew it was his way of saying, "We're going to be good friends."

And, indeed, Sannado was right. We did become good friends. Almost daily, I would take an old garbage can lid and walk out into the arena. When hit with a substantial stick, the lid made an excellent drum and a drumbeat was what Sannado loved most. Upon seeing me enter the arena carrying the "drum," this splendid horse would immediately station himself solidly on all four feet and wait. The first beat of the drum was the trigger to move this young colt into a performance of which the utmost of critics would approve.

This Toronado son had a keen sense of timing. He pranced in time with the beat without missing one. He would change the cadence of his trot to fit that of the drum. Sannado would often flare his nostrils and snort as if to say, "Challenge me. Challenge me," at which point, I sometimes got creative just to show him, "You are not quite the cat's meow that you think you are." But he really was. This wonderful horse was never more beautiful than when he was doing a skip trot to the changing beat of the drum.

Sannado was still too young to ride but, one weekend, Rheta and some friends hauled him to Lubbock, Texas to an Arabian horse show. He took second place in a field of twelve in the Liberty class. Liberty is where a horse is turned into the arena accompanied by its handler but can "do as it chooses."

This practice might prove embarrassing to some handlers, but Arabian horses do not need help showing off. They do it quite well on their own. I almost think Sannado understood his accomplishment. Upon returning home, the now bona fide show horse got extra treats. The next day found him eagerly waiting for playtime.

Breaking horses is not like you see in movies, nor is it the way it was done long ago in the Old West. Back then, cowboys had no time to waste prepping broncs to accept a saddle on their back. They just threw one on and climbed aboard. The horse would buck until it either threw his rider to the ground or he gave up and accepted his fate.

I still remember the day that my brother-in-law, James Golightly, and I put a saddle on his gelding, Scout, for the first time. I put my foot in the stirrup and eased myself into the saddle very, very gently. Scout just sat there until I kicked him in the side, at which time we had ourselves a rodeo.

When Scout stopped bucking, I was barely hanging on. After that, he never bucked again, and, for more than two decades, many nephews and nieces put that horse to good use when they visited Nita and James.

My first attempt to break Sannado to ride was a bit of a different experience. I snapped a lunge line onto his halter and ran him around and around in a circle to use up some of his pent-up energy. Then I fitted a bit in his mouth and pulled the bridle over his ears. Using a saddle blanket, I gently rubbed him with the blanket until he was no longer afraid of it. Next, I let the blanket flap against his flanks until he accepted that. Then I placed the blanket on his back and carefully set the saddle on top of the blanket and cautiously cinched it. When I was sure Sannado was ready, I walked him around and around to let him get a feel for the saddle.

Once Sannado accepted the saddle, the next step was to get him used to the pull of the bit in his mouth. This step was the risky one. I unsnapped the lunge line from the halter and hooked it to the bridle. I, then, tied the reins loosely to the saddle horn and was ready for the next step. Gradually, I pulled his head around slowly, pulling then releasing, pulling then releasing, until I thought he was ready.

It has been said, "Hell hath no fury like a woman scorned." I beg to differ.

When I commanded "walk," and gently slapped him on his backside, that son of Toronado became a tornado. He reared straight up and began falling back. My heart leaped up into my throat. Not only was I going to destroy my saddle, but my horse was going to break his neck or back. He did not though. He promptly got back on his feet and gave me the look. "How could you do this to me? I thought we were friends."

After getting him used to the bit, Sannado agreed that we were friends and he allowed me to ease myself into the saddle. He never bucked. A friend of mine often tells folks, "Arabians don't know how to buck."

Within thirty minutes I had this magnificent rose-colored stallion moving smoothly around the arena's perimeter until the wind blew a large piece of corrugated steel into the fence. Sannado went straight up and so did I. He came down first and I followed, striking my tail bone on the back of my saddle. Ouch. It hurt to sit on a hard surface for months.

By the way. If you have never had a horse for a friend, get one. Now!

Hound Sounds

Many was the night, as a young boy, I laid in my bed listening to sounds that now, so many years later, have become dear to me. I long to hear them now. Something about them was, I think, orchestrated by the Creator. I am speaking of the sounds of hounds.

I always thought that my first exposure to an orchestra was when I was an airman stationed at Lackland Air Force Base and was forced to attend a performance of The San Antonio Symphony after which I changed my mind about classical music. I liked it. I liked it a lot.

My sudden change in musical taste was hard for my sister Nadine to understand since many times I had laughed at her for liking "The Saber Dance." The only "real" music, in my opinion was hillbilly music.

But, I was wrong. My first ever orchestra to hear was the performance of the fox hounds. Uncle Art Parr and his friends would turn their fox dogs loose on the ridge south of our house and it would not be long until they would open in unison and perform a remarkable symphony. One hound on a fresh fox scent is a thing to hear but let that one be joined by ten or twelve others and, friend, you have yourself a real fine musical production.

I have always favored the barking of hounds to that of any other breed. Some dogs, to me, have very annoying barks, especially small dogs. Some sound mean. Unpleasant. Not hounds though. I have never heard a hound howling I did not like. Just as an orchestra has many instruments combining to produce a variety of sounds, so do hounds.

One dog might have a high-pitched voice while others give out low guttural strains and all ranges in between. Sometimes you can imagine you are hearing a sonata. Listen carefully and you can hear the exposition of their theme as they first pick up a fox's scent. Their excitement increases, and the theme's development begins. The recapitulation signals that the fox has either been caught or chased to its den.

I can't say when fox hunting went out of style, but it did. I only remember hearing the hounds run when I was quite young. Perhaps it was because Dad had downsized to a house with five acres, not large enough to accommodate the hunt. I have tinnitus and the sounds I hear when things are quiet are those of the night. I hear crickets and tree frogs, but none are there. I enjoy my affliction because I love night sounds. I just wish that once again I could hear the hounds run. But that would mean some poor little fox is being tormented so I think it best that I just remember the hound sounds.

Of Crocheting, Quilting, Knitting, and The Like

I was about ten years old when my mom taught me how to crochet. She was good. She spent hours making beautiful doilies. They were the ones with elaborate patterns that are heavily starched so that all the frills stand up to make gorgeous end table décor. As I recall, we had lots of end tables and a couple of "what-not" shelves that were placed in room corners. With my newly acquired crocheting skill, I only made one thing, a belt.

But that was long ago and would likely be forgotten were it not for Rheta liking to crochet, knit, and weave. When she picks up her current project and starts her work, I often remember my mother seated in her rocking chair with the needles turning out beautiful cloth pieces. I wonder whatever happened to my one and only product of my crocheting skill.

Candles, Kerosene Lamps, and Electricity

Our kids do not know much about the early life of their parents, that we studied by coal oil lamps. Perhaps this brief discussion will lend some measure of cause for them to better appreciate life's comforts today.

We had no electricity until I was eleven years old. So, we kept candles and matches at hand on the dressers in our bedrooms. It is true that we had flashlights, but the candles provided extra insurance should it be that we reached for a flashlight and found the batteries had run down. We had no indoor plumbing, so we made frequent use of flashlights when we had to pay a visit to "a man about a dog." (Our way of saying, "I have to go to the outhouse.")

Two or three kerosene lamps were present in the living room, dining room, and kitchen. Most of the lamps were just plain vanilla lamps. They had an oblate shaped glass globe that tapered a bit at the bottom to fit onto the base of the lamp and the diameter decreased at the top. A cloth wick extended from the burner down into the bowl of kerosene that comprised the base of the lamp. A control much like the winding stem of a watch facilitated the increasing or decreasing of the flame to increase or decrease the light. When it was time for lights out, Momma hovered over the globe and gave a quick puff of air through her lips to extinguish the flame.

We had one "Aladdin" lamp which had a more beautiful base and taller globe. The taller globe gave a brighter light, and it was the lamp we used for doing our homework.

Daily cleaning of the lamp globes was essential, as was the trimming of the wick That job fell to us kids. We also kept a lamp in the root cellar where we took refuge from storms.

When I was eleven, we were excited to see a big REA (Rural Electrification Administration) truck coming down our private road. Electrification of rural Oklahoma was life-changing. It meant that soon there would be light in every room. And, we would not have to clean lamp globes nearly as often. It meant that dad would have the benefit of a skill saw rather than the hand saw he used to build our house. I still remember the big auger on the truck as it drilled a hole for the pole. The pole was set. The wire that would soon carry the magic current was strung. The fuse box was set. The lever on the box was pushed up to complete a circuit and, suddenly, there was light. I remember those days with fondness, the days of candles, kerosene lamps, and electricity.

Where Panthers Scream and Mad Dogs Roam

World War Two ended on the second day of September nineteen-forty-five, exactly six years and one day after it had begun. After that, we children no longer feared that the Germans and Japanese would take over America. Life was good. Well, at least, mostly for there can always be found something to fear. In our case, it was panthers and rabid dogs. Panthers (I now know them to have been cougars) roamed the ridge just south of our house and in the mountains only a few miles to our north.

The mode of travel for young male suitors back then was the horse. Much of dating consisted of the boy riding his horse to see his love interest. Occasionally, young men of our community, my brother included, would tell of having been followed by a panther. They would talk of snapping twigs and rustling leaves that excited their horse so that it would sometimes give out with a snort. Such news always provided subject matter for late evening conversations on our front porch. Sometimes friends or other family members would come over on summer evenings and add to our discussions of such topics. Summers were hot and daylight savings time meant that it would be late before we could retire for the night and not suffer too much from the heat. We were also out of school and could stay up as late as our parents. After listening to the adults talk about panthers, our minds turned the one into a thousand.

On another occasion, my brother in law, James, went out to get a bucket of coal. The coal pile was next to the garage. He turned on the headlights of his car to see. When he did, what he saw struck fear in his heart. Two scarlet eyes just a few yards away. He quickly filled the bucket and came back into the house. Did he tell that story for my benefit? I will never know but it sure got my attention.

Many years later would find that wide-eyed little boy married and living in Texas with a wife and two children. We built a new house on fifty-six acres south of Springtown. Some nights in the summer we opened our bedroom window on the second floor of our house. I loved the night sounds. But sometimes we could hear more than crickets and tree frogs. I had always heard that a panther scream sounds exactly like a woman crying out for help. The first time I was able to verify the truth of that claim, I was suddenly awakened from a deep sleep. My first thought was to grab my gun and a light and go out to help the woman who was crying loudly, "Help. Help" but dragging out the words help. It was such a mournful sound. After listening for a few minutes, I remembered. It was a cougar.

Succeeding years would prove my suspicion that it was a "panther." Several of my sister in law's goats were taken and claw marks were left on the back of a cow. Today, we shudder to think that we allowed our kids to play near the scary woods.

Yes, fears of panthers and mad dogs had replaced those of Germans and Japanese. Did I say mad dogs? In the nineteen-forties, folks did not vaccinate their pets for rabies. In fact, they did not even worm them except occasionally someone would advise to give the dog a copper penny embedded in a biscuit. I had no clue as to whether it worked or not. Now, I feel pretty sure it did not. Dogs would get what was called "stiff fits" and "running fits." A stiff fit was where the dog would lie down and stiffen its limbs and shake. A running fit was as the name suggests. The dog would run and run, often in circles. Worms were the cause.

On rare occasions though, word would get out that someone had spotted a rabid dog. That was a horse of a different color. Or, should I say a dog of a different affliction. News like that would put the whole community on alert. Nobody would go outside without being aware of their surroundings. The road that stretched a quarter of a mile from our house to the number five road became a more frightening thing. Normally, I would only be anxious when I rode my bike alone at night on my way to a basketball game at the high school. I was always sure the ravine that ran perpendicular to the road was full of demons. Trolls, panthers, and now even mad dogs lived there.

A boy's wild imagination usually merited no consideration. But, when actual sightings of a rabid dog were reported, it was a different story. Twice, as a youngster, I had dogs that had to be destroyed because it was thought they had been bitten by a rabid dog. Whether the bite was rabid or not was immaterial. I recall hearing a dog fight just outside my window one night. The next day our dog Cubby had a big wound on his neck. My mother walked to our neighbor's house asked him if he would put the dog down for us. Dad did not have the heart to do it and our neighbor was kind enough to take care of it for us.

Mother said that when she walked out into the woods with Mr. Dill and Cubby, our beloved pet stood right next to her so that Mr. Dill found it hard to get off the shot that would end the live of my pet. But he did. Shep was another of our dogs to become victimized by a rabid dog. Shep was put into a corn crib in the barn and I presumed he would remain there until he either went mad or it was safe to let him out. One day I went to look through the cracks to see my dog, but he was not there. I cried because I knew I would not see him again.

Nowadays, we do not worry so much about rabid dogs although folks in rural areas still must be concerned if they see a skunk during the day. It is very unusual to see a skunk in the light of day because their nature dictates a nocturnal life. I suppose we have come full circle since the days of yore when Germans and Japanese were threatening to conquer us followed by fears of rabid dogs. Now we are back to being concerned that war once again looms a possibility what with Russia aiding and abetting Iran, a major exporter of terrorism. North Korea is spoiling for a fight and ISIS wants to inflict harm upon us. I wonder, will we go back to the days when panthers scream and mad dogs roam? I hope not.

Lost in Peachtree Hollow

The San Bois mountain range is a narrow strip of small mountains about forty-five by twenty miles. Their height is a little over eighteen-hundred feet. Though small, these mountains are very rugged. The slopes are mostly very steep and there are many cliffs spread throughout. The San Bois which we just called "the mountains" run west to east through the little community of Panola, Oklahoma where I was born and grew up. They played an important role in the life of residents in Panola and surrounding communities in the nineteen-forties and fifties. Wildlife has always been plentiful there and many a table was well supplied with venison and squirrel. Now, even black bears can be harvested.

Oklahoma became a state in nineteen-aught-seven and my dad W. T. Mankin, an eight-year-old boy, was there to see it. He grew up hunting in the San Bois to help feed a family of eleven. There were no hunting seasons. I can't say when was the first time there was a deer season requiring a hunter to have a hunting license, but Dad always felt that since he was in Oklahoma when it was still Indian Territory, the hunting laws did not apply to him. He did not change his mind about that until a game warden gave him a ticket for hunting out of season. That got his attention when he had to appear before a county judge and pay a fine. When he objected the judge, his friend, said "Willy, if you get caught again you will not only pay a fine, but I will confiscate your gun. As far as I know Dad never again hunted out of season.

When I got old enough to deer hunt, Dad would take me with him to the mountains. Henry had a cabin on the bank of Little Fourche Maline Creek. It was just one room with a stove, fireplace, and bunk beds. To the north of the cabin was Yancy Mountain, one of the tallest in the range. Deer hunting season opened in November. Novembers in southeast Oklahoma can be quite cold, and often, in late November, rain or snow or a mixture of both was always a possibility. I remember once on a hunting trip, we had just retired for the night in our warm dry cabin when we heard a rifle shot. Moments later we heard another shot. Since it was well after dark, the rifle shot could mean only one thing. A hunter was lost. My brother and a friend, Herman, got a couple of raincoats which were kept in the cabin and headed out on horseback, to go to the sound of the gun shot. After a while they returned with the poor, cold, and wet hunter. I will never forget how happy he was when we invited him to spend the night and have a hot breakfast the next morning. Little did I know then that some years later, I too would be lost in the same mountains.

Peachtree Hollow is a favored spot to find deer but was also a favored spot to get disoriented and sure as shooting, I did. It was late afternoon and I had decided to head back to the cabin. That morning I had crossed Little Fourche at a shallow ford with rocks that could be used as stepping stones. I had hunted Yancy but without being aware had dropped down into Peachtree Hollow. I had always heard that when you get lost, you keep walking in circles, passing the same spot several times. I am here to vouch for the truth of that saying. As the Cajun comedian Justin Wilson would say, "I was so lost, I didn't know where I was lost from and that's bad." Dusk was setting in. I had to find my way back before it got too dark to see.

I started walking west along what I thought to be the south side of Yancy. While it was still light I saw that the creek was deep and lined with thick brush, the kind that would scratch your face bad and even poke you in the eyes if you did not exercise great care. I decided it was time to fire off a round with my thirty-ought six rifle. Moments later, I heard a rifle fire west of me. I was on the right path. I continued to walk in that direction. I was too far from the cabin. In what seemed to be mere moments, I was stumbling along in dusk dark. After firing all but one round, I knew I had to change plans.

Realizing that darkness and walking along creeks are a bad idea, I began gathering wood with which to fuel a fire. The fire would serve a dual-purpose. It would reveal my location if my dad and brother got close enough to see and it would keep me warm in the night if they did not. Fortunately, I had remembered what Dad had always drilled into my head. Take lots of matches and keep them dry. I had. The fire was a good one. I was proud of myself. The weather was good, unlike when the other hunter was lost. And, the ground was dry and now warmed by the fire. I laid myself down tired and ready to sleep.

A short time later I heard music, the kind that causes the heart of a lost hunter to leap like the deer he hunted. It was the sound of my brother's jeep coming up the mountainside. I instantly leaped to my feet and began yelling, "can you see my fire?" That was a huge mistake. Not only did I have to endure listening to my dad and brother teasing me, but Dad never let me live that down. He always managed to somehow work into our conversations "do you remember…" and then he would laugh and say, "can you see my fire." Dad has been gone since nineteen-eighty-two. So often I have reflected upon those times when we walked the mountains together. Oh, how I wish I could again be lost in Peachtree Hollow and hear my daddy say, "Yes, I can you see your fire."

For Nadine

She was born in nineteen-thirty-four, the last-born daughter of W.T. and Birdie Mankin. They named her Nelta Nadine. As a sidebar, it is interesting to me that all the children born to Birdie were called by their middle name. Because of only two years difference in our ages, Nadine and I were closest. We were buddies. She was my heroine. I suppose also that perhaps I was her hero. I shall explain.

When we were young, we were inseparable. In grade school, she learned to read before I did, and she would read to me. We wanted the best for each other. Back then, we always had two catalogs lying around. There was a "Sears and Roebuck" catalog and a "Montgomery Ward" catalog. Each year, new ones would arrive in the mail. The old ones were placed in the outhouse for toilet paper. Yes, I said toilet paper. Just do not use the slick pages.

These catalogs were called "wish books" by many. They certainly were for us. Often, we could be found sitting on the divan or, as it is called now, the couch, turning pages until something caught our eye that was mutually desired. Each time that happened, we pointed to it and said, "one for you and one for me." Pages were turned until, again, we came to something we wanted. Once again, we claimed it with the declaration, "one for you and one for me." We were the richest kids in the world. Looking back on those days, I venture to say that we had more nice clothes and toys than the Rockefeller children.

Nadine taught me to climb trees. And, I might add, she

taught me how to find my way back safely to the ground. She was a tree climber extraordinaire. She could skinny up the tallest of trees. I could not. She would point to the limbs that I could reach which would support my weight. Most importantly, when we got as high up the tree as we dared go, I would freeze, unable to convince myself to climb back down. I am still afraid of heights. Nadine would identify the limbs I should use and offer lots of encouragement. Moments later, I would be back on terra firma thanks to her.

I was a skinny kid and older boys would pick on me but not when Nadine was around. Looking back on those days I now realize that I had my own special body guard for free. Once, I was riding Snip on the school grounds during the summer and a boy in the neighborhood began throwing rocks at me. Nadine was with me on her bike. She got off her bike and lit into him. He fled. Once again, my heroine rescued me.

Yes, Nadine was my protector. But, in my own way, I was her hero. Nadine could not swim as well as I. So, once after prolonged heavy rains, Little Fourche Maline Creek where we went to swim, had swollen to near bank-full. Nadine ventured too close to the swift water and was in danger of drowning. I managed to get close enough to reach out and grab her hand. Together we got back to dry ground.

Another time, we were playing on the roof of the well house and Nadine fell off. She was knocked unconscious. The only telephone in our community was the one that belonged to Mr. and Mrs. Garner who owned the local gas station, grocery store, and filling station. Momma instructed me to go ask Mr. Garner to call the doctor. Dr. Henry's office was in Wilburton, about seven miles away. Of course, when Dr. Henry arrived, Nadine had regained consciousness, and all was well. As I recall, he accepted jars of fruit and vegetables that she had canned. Bartering for such things was common then.

I recall almost losing my sister and heroine once when she

came down with "tick fever." She had suddenly broken out in a rash with an attendant elevated temperature. Ole Dr. Henry knew his stuff for that day and age. He diagnosed "tick fever" and began treatment. For a while, it was touch and go and it even appeared we might lose Nadine. Ice bags and an oxygen tent were used and eventually her fever returned to normal.

I later learned that Momma had told Nadine that when she was near death, she had told God that she would be willing to give Nadine up if her life would not honor God. At that very moment, Nadine opened her eyes and recognized Dad. Although many years have gone by, I still remember those days with fondness and wanted to say, "Thank you, Sis, for being my heroine."

Long, Long Road to Alaska

The Alaska-Canadian Highway is paved today. Such was not the case when we traveled it in nineteen-seventy-seven. Driving two vehicles, our family set out for Anchorage where I would begin a new assignment. It was to be the mother of all memories.

For two years, I had been teaching fundamental principles of meteorology to Federal Aviation Administration employees at the Academy in Oklahoma City. Though not a part of the National Weather Service, several years prior to my teaching assignment there, The National Weather Service had turned over pilot briefing responsibilities to the Federal Aviation Administration. Only a rudimentary knowledge of meteorology was needed to advise pilots of weather along a route. We, the National Weather Service, still provided the aviation weather forecasts.

At the Federal Aviation Administration Academy, I was a GS-twelve and had been offered a GS-thirteen position at Anchorage, Alaska. It would be an exciting job. I had studied tropical meteorology and was able to use that knowledge as a forecaster in San Juan, Puerto Rico. Now, I was going to have an opposite extreme experience in Anchorage. Since the atmosphere is an ocean of air, it behaves in compliance with fluid dynamics and because of things like air density, pressure, angular momentum, and such. Weather patterns are vastly different near the equator and poles than in the mid latitudes.

Indeed, I knew I was going to meet new professional challenges in Alaska and I was chomping at the bit to get there. A few days before we loaded up our Toyota and Ford station wagons and headed out of El Reno, Oklahoma, I had bought new tires with studs for each station wagon and had one freeze plug heater installed on each. We would soon find out the value of those things.

We left El Reno on December twenty-third, bound for Alaska with two kids, a whippet named Wendy, and a white cat named Snowflake. Oh, if we had only known! We had sent in a request to Texaco for a trip kit. In the seventies, before personal computers, oil companies would map out your travel routes free of charge. Ours took us through Oakley, Kansas, Cheyenne, Wyoming, and Sheridan, Wyoming, where in each place, our patience would be tried, to the limit, and we would fail the test.

Early on the morning of the twenty-third, we backed out of our driveway and began our journey. Carrie rode with Rheta in the Toyota Corolla station wagon and Steede rode with me in the Ford. As I think about it now, our trip was somewhat marred from the get-go. Rheta's brother Arla Cole had given me a pair of Mille Fleur chickens which I kept in a coup. In my excitement to get on the road, I forgot to arrange for someone to take them. You know the rest of that story. I felt bad, very bad.

The next crisis involved Wendy the whippet. The dog was given to Rheta by a man with whom she worked. She was a beautiful animal, but it did not take long for us to realize what a terrible mistake we had made. I do not think that one could have found more methane gas anywhere in the world. That dog had it all pent up in her skinny little frame and was very generous in sharing it. I think if the temperature outside had been sixty below we would still have ridden with the windows down. When we had taken as much as we could take of the foulest of odors Steede and I traded off, the dog for Snowflake the cat.

We spent our first night in Oakley, Kansas. What a night that was! We knew we could not leave the two animals unattended loose in the motel room, so we tied Wendy to the bedpost and left Snowflake free to get to the litter box in the bathroom. That seemed a good plan to us. It was not. When we returned to our room we were met with a shredded bedspread on our bed. "That stupid cat," I exclaimed. Now I had to go, hat in hand, and apologize to the motel manager. We paid him the sum he required and headed on down the road. After we discussed what happened, we concluded that it was not the cat at all. We figured the cat could not have torn such a large area of the bedspread. Clawing would have probably been confined to a section of the spread near the floor. When we got to Cheyenne, our next sleepover, we knew what we would do.

We would put Wendy in the bathroom and close the door. That would take care of it. I was proud of my problem-solving skills but should not have been. We left Snowflake in her crate. Upon returning from dinner, we found the carpet near the bathroom door destroyed. An inch or so of carpet extended under the door to the bathroom and Wendy went to work on it. She had scratched all along the space between the bottom of the door and the carpet. It was shredded. Embarrassed, once more, I went to the manager to pay for the damage. He was very understanding and told us he had enough left-over carpet to repair it. He only charged us twenty dollars. We thanked him profusely and got out of there quick. It did not take us long to decide that when we got to Sheridan we would take Wendy to the Humane Society. We got to there in time to drop Wendy off at the Humane Society. We gave a donation and left the sack of dog food we had. We even had time to celebrate Christmas by seeing a movie.

The next morning, we departed Sheridan and headed for our next stop, Great Falls. The journey from Sheridan to Great Falls was uneventful if you do not count the beauty of hundreds of pronghorn antelope running across the plains. We departed Great Falls early the next morning. North of Great Falls, the well paved highway petered out and became a dirt road. Recent rains had left potholes full of water and mud which coated the lens of our headlamps and made us think our lights had burned out. We could hardly see the road. I cleaned the headlights with Windex and paper towels. We could see again. Off we went.

Our next stop would be Calgary. When we reached the Canadian border, it was daylight. Before entering Canada, we had to declare whether we had any firearms. Yes, we did have. My Mauser three eighty hand gun was in my vehicle. It was not loaded and when I showed the authorities my transfer documents, they told us to go ahead but to strictly refrain from using it while in Canada. Now that was a comforting thought. What if we met up with a bear. Oh, wait, well now, how much would a little three eighty worry a brown bear? Not at all I imagine. We would not dare use it. I bet the Customs guy still laughs about the Okie who thought he could kill a grizzly with a nine-millimeter short (three eighty).

We crossed into Canada and within minutes I called Rheta on our CB radios. "Gypsy (her handle), you got your ears on? Come back."

"That is a Roger, Prospector (my handle)."

"Gypsy, I just saw the ugliest horse and colt ever. Wait! That is not a horse. It's a moose."

We both got excited. We had seen our first moose. We would see many in the next four years. Occasionally, Steede and Carrie would take to the airways. Their CB handles were "Little Cottontail" for Carrie and "Pinball Wizard" for Steede. You would have thought we were truckers, "come back." Shortly after we saw the moose cow and calf, we saw an Arctic fox and snowshoe rabbit (not together of course).

We arrived at Calgary in time to do laundry that had accumulated. We liked the town. I suppose the fact that it is a sister city to Fort Worth had something to do with our affection for it. The part of Canada around Calgary and Edmonton is as flat as a pancake. Calgary is, of course, the home of the Calgary Stampede. They have an annual rodeo and stock show but it was the wrong time of the year for it.

We stopped in Edmonton and bought down jackets for each of us. We got some good ones at a shopping mall there. From that point on in December, it was mighty cold. Twenty-five to thirty degrees below zero was going to be a reality the rest of the winter.

Leaving Edmonton, we headed on toward Anchorage. So far, except for an unruly dog, the trip had been exciting. We had seen animals unlike those in Oklahoma and had experienced vast stretches of prairies. What we were about to see would have made us want to be transported there immediately but what we would soon experience would have made us want to turn around and go back to Oklahoma. We did not know so we went on. We kept on trucking over hills and prairies.

We took highway forty-three out of Edmonton. It took us through several small towns. Nothing noteworthy was happening in any of the little towns, so we just kept driving, getting out occasionally to stretch. Each time we did get out though Carrie, who was seven, had to crawl out from among all the "stuff" we had packed around her. And, we did not stretch for more than a minute or two because the temperature was well below zero. And, as for the stuff we packed around her, well it would come into play in a most unusual way on down the road.

Shortly before we arrived at Dawson Creek, the fan motor went out on my Ford station wagon. Steede and I donned our down jackets and gloves and continued. Our destination for the day was Fort St. John but we ended up spending several hours getting the heater fan replaced. We sure were glad to get that heater going again. We continued toward Fort St. John. Arriving there, we pulled up to the motel electric rail and plugged in our cars. The temperatures were so cold at night that it was necessary to keep the engine warm. Otherwise, the car would not start the next morning. As I plugged our cars in, I was reminded of those old Roy Rogers movies where they would ride up to a place and tie their horses to a rail.

We awakened the next morning and began to get dressed. Rheta went around to the motel's restaurant to get us some coffee and hot chocolate for the Steede and Carrie.

In a few minutes, she came bursting into the room, announcing, "I saw a Mountie. I saw a Mountie."

Later, on our journey, the Royal Canadian Mounted Police would endear themselves to us in a matter of grave importance.

We had breakfast at the motel restaurant, and then headed out on highway ninety-seven. The highway was not bad if you can deal with occasional frost heaves. It was only two lanes but the chance of meeting a car coming from the opposite direction were quite slim back then. The distance from Fort St. John to Fort Nelson was only two hundred thirty-seven miles but what with snow covering the highway and snow falling it was a slow go, a scary slow go. And we would be the only ones on that stretch of road with literally no towns in between. Civilization was non-existent. The studded tires worked well but we would soon run out of prairies and get into the mountains. Then we would have to dig out the chains. Twenty-five to thirty degrees below zero weather is no time to be running off the road. Little did we know that running off the road in subzero weather would be the least of our worries.

We were almost a full day getting to Fort Nelson and we were road weary. Driving on icy roads raises one's stress levels to the limit. We tied our horses up, I mean we plugged our cars in, and checked into our room. As we began taking our nightwear out of the suitcases and preparing for bed, I noticed that we had not brought my briefcase in. I returned to my car to fetch it since I had a check for no small amount in it. The money was an advance on per diem and had to pay for food and temporary quarters until we could get settled in Anchorage. The briefcase was not there.

With a knot in my stomach and a lump in my throat I started discussions with Rheta about our options. Should we turn around and drive the snow-covered road back to Fort St. John? Since the sun had set at two o'clock, it was well after dark. We would first explore other options. We called the Royal Canadian Mounted Police and they told us that they would send an officer over to the motel in St. John and if the briefcase was there they would secure it and mail it to us. We called the motel where we had stayed, and they verified that, indeed, the briefcase was there. It was locked and so we breathed a sigh of relief. When we awoke the next morning, we had a good breakfast and got back on the road minus my briefcase.

Shortly after leaving Ft. Nelson, we began gaining in elevation. The Canadian Rocky Mountains loomed majestic ahead. Such beauty we had never seen. A few small streams in Alberta are fed by hot springs. The unfrozen creeks and very cold temperatures team up to provide some of the most remarkable scenery. The vapor pressure flux from the warm open water of the streams, upward into the extremely cold and dry air causes the formation of hoar frost. Find the most beautiful Christmas postcard you can and then imagine that scene multiplied over the vast expanse of Canada. This, we experienced again and again. Black spruce is common in Alberta. When covered with hoar frost and bathed in sun rays, the sight of those trees is one to behold.

Soon we would be wishing to be back in Fort Nelson, back in Fort St. John, back in Dawson Creek, back in Edmonton, back in Calgary, indeed, back in Oklahoma. But we were not. So, on we would go at a snail's pace with the dread of terror just around next curve. Terror? Why? What we were about to experience was not even in our imagination of how things can be in the Canadian Rockies. When the ALCAN highway was constructed in nineteen-forty-two, I do not think they had even heard of guard rails. As I mentioned before, the highway to Alaska is paved now. It was not when we traveled it. It was gravel and we were entering some of the most rugged mountains in North America.

As we began to ascend the Canadian Rockies, we made more use of our CB radios. Sometimes, we just wanted to point out marvelous sights and talk about them. The deer and Dahl sheep gave us reason to occupy the Citizen Band.

"Gypsy, you got your ears on? Come back. Look on your left. See the moose and calf in the bog ahead?"

Gypsy continued, "Prospector, you got your ears on? Come back."

"Yes, Gypsy, come back."

"Prospector, I just saw an Arctic fox run across the road."

"Gypsy, I can go you one better than that. I just saw a Dahl Sheep on that cliff I just passed. You should be able to see it now. Are you okay back there?"

We found the CB radios indispensable for another reason. "Breaker, breaker, Gypsy, you got your ears on, come back?"

"I am hearing you, Prospector."

"Gypsy, I am thinking we should stop at the next flat place and put on our chains. That mountain coming up looks rough."

"Roger that, Prospector."

The roads were snow covered, sometimes packed and icy. Chains were essential. Once firmly attached to the tires, we felt a bit easier about the road ahead. We soon realized that, while we were in the mountains, easy feelings would be as scarce as hen's teeth.

The main reason for the lack of peace about our circumstance took the form of sheer cliffs on our left with no guard rails. As a weather observer, I had estimated the height of cloud bases many times. I had little experience estimating the distance of cliff heights looking down. I tried to guesstimate how far down it was to the bottom if, God forbid, we should plunge off the road. We hugged the cliffs on our right a little tighter, trying to gain a bit more badly needed courage to go on. I estimated several hundred feet for most and perhaps near a thousand feet for others. No matter though. Should the unthinkable happen, survival would be impossible.

Mile five fifty-four on the Alaska Canadian Highway is a number that will forever, apologies to President Roosevelt, live in infamy. Early on the morning of December thirty first, we had left Fort Nelson and had reached the mountains an hour or two later. We had tire chains for both cars but only used them when we absolutely had to. Chains slow you down and chew up the tires if left on at highway speeds. So numerous times, we would stop, remove them, drive on, stop, replace them and drive on. With bitter cold awaiting me outside the car, putting on and taking off chains was a real patience-testing activity. Rheta and the kids did not mind at all. Well, I must be honest. I did have their sympathy.

We now found ourselves high up in the mountains, and I do mean high up. What we had called mountains in Oklahoma would never be so called again. When Jimmy Rogers penned "In Those Oklahoma Hills Where I Was Born," he knew the difference between hills and mountains. I wonder if he had travelled the ALCAN highway. Call them whatever, the fact was, there we were, and we had to make the best of it. Our car heaters worked double time trying to keep us warm, but I do not remember ever being snugly warm like I wanted to be. A gigantic mountain peak came into view ahead. So out I got once more and put the chains back on.

Of course, mountain roads wind of necessity. Mountain roads wind around and form switchbacks. The reason for this is to avoid ending up with the road being too steep. A winding road with switchbacks affords a much safer journey up a mountain. Also, you get where you are going without getting so high that you run out of oxygen. At any rate, if my memory serves me correctly, this mountain was called Steamboat Mountain (not to be confused with the one in Colorado).

Up the mountain we went although at a somewhat fear induced speed. Slow. The last thing we wanted was to run off down the side of that mountain. Stress levels rose like the peaks ahead. The going down was much worse than the going up. Did I say much worse? Sheer panic at times, but quiet panic because I certainly did not want to convey to the children that Daddy was a coward. Rheta was such a trooper. She took it in stride as though she had done it a hundred times. Let first gear be your brakes. The conditions themselves forced that firmly into our noggins. Keep that foot off the brakes when you are on ice. Steer into the direction of your slide.

Finally, the top of Steamboat Mountain. The sight could not have looked better to us. A little gas station, grocery store, café combination. Hot coffee! Hot diggity dog! Get me to the restroom. My bladder is about to break. Oh, no restrooms? Please tell me it ain't so. Oh, it's outside? Point me in the right direction. Thank goodness, it is only a short distance. Will these Huskies bite? They sure are pretty. Steede and I found the men's and Rheta and Carrie found the women's. Snow was deep. Carrie was only six years old so Rheta had to practically carry her. The old dogs were following. I think they wanted to watch the entertainment about to happen.

We reached the outhouse not a second too soon. Relieved, Steede and I stepped out and began petting the dogs. Rheta followed two or three minutes later. It happened that in both outhouses the seats were ice covered and filthy. None dared to sit. Rheta lifted Carrie and tried to position her the best she could. Full bladders wait for nobody, certainly not for precise positioning over the hole. The result? Wet jeans on Carrie. In thirty below zero, liquids freeze instantly. So, it was not a problem until the warmth inside the car thawed them out. The smell of urine accompanied Rheta and Carrie to our next stop.

One can drive two or three-hundred miles in those parts and not see a gas station. We tanked up on gas and coffee while on top of the mountain. We also had lunch and then hit the road again. We reached a plateau where only gentle slopes greeted us for a while. The landscapes were amazing. Fields of deep snow on the right. We dare not look out the driver side windows. Cliffs several hundred feet straight down were on that side? Big snow plows had scraped the roads during the night so that driving was not bad if you did not focus on the question, "What's it like to run off a cliff and die?" The plows had left show berms two to three-feet high and I had just gotten my right front wheel in one and almost lost control.

"Gypsy, you got your ears on? Come back."

"That's a big ten four, Prospector."

"Pay attention to those snow berms on the side of the road. If you get a wheel in one you might roll over. Come back."

"Don, God didn't bring me all this way to put my butt in a snow bank. Come back."

"Roger that, Gypsy. Just be careful."

Sometime before reaching mile five-twenty-four, we started seeing signs announcing Fireside Lodge at that marker. We had planned to try to make it into Watson Lake to spend the night, but we were tired, and we liked the sound of bedding down at Fireside Lodge. As we arrived at the lodge, we saw a sign. "Closed for the winter" it said. Disappointed, we plugged on toward Watson Lake still discussing what God would or would not do with respect to our wheels and those snow berms.

"Gypsy, just remember that God tested Job in strange ways, come back."

But she did not come back.

"Gypsy, you got your ears on? Come back. Breaker for Gypsy. Gypsy, are you hearing me?"

I thought perhaps the mountains had rendered a dead spot and she would be out of it shortly. I tried again. No answer. I found a place to turn around. Why is she not answering me? As I rounded a sharp curve in the road that I had just passed a few minutes earlier, I saw the reason for her silence. Her CB antenna on the top of her station wagon was now pointed toward the center of the earth. She was upside down. All four wheels of her vehicle were pointed skyward and still spinning. My heart was pounding as never before. It was the most awful kind of fear I can imagine. I knew that she and Carrie were dead since they were nowhere to be seen. Then I breathed a half sigh of relief. Rheta had been trying to kick the door open. When the station wagon rolled, Rheta had instinctively turned off the ignition and then began trying to open the door. She could not. Snow had packed around the vehicle. It was deep.

As I pulled off to the side with my window down shouting, "Are you okay? Where is Carrie?"

"We are okay. Carrie is in the back and okay."

We exited our station wagon and trudged our way through deep snow to where Rheta was helping Carrie get out from among the "stuff" we had packed around her. According to Rheta, as the station wagon went upside down, Carrie had cried out, "Mommy, things are falling on me." Rheta had leaned over to wipe a fogged right front window. A big mistake. She got her right front tire in the snow berm and, indeed, God had brought her "all the way up there to put her butt in a snow bank." He had used the law of conservation of momentum to do it. In other words, when the right side of her vehicle was suddenly slowed by the deep packed snow, the amount of momentum lost to the right side was gained by the left side. This action supplied a sudden right-directed torque and over she went. God was merciful though. He could have put her over the unguarded cliff but chose instead to turn her station wagon into a sled, giving her a ride over a snow-covered field. That day we praised God for His infinite mercies. We still do.

After a few minutes, the shock diminished and reality set in. What now? It was two o'clock in the afternoon. The sun was setting and here we are. At twenty-six below zero, we could not tarry. We had to decide quickly. It was eighty miles to Watson Lake but only eleven miles back to Fireside Lodge. We decided to take as much from Rheta's station wagon as we could and leave the Toyota Corolla sitting there. We hoped to find someone there who could pull us out once we got to Fireside. We were just about to start removing "stuff" from the Toyota when a guy in a pickup stopped. Normally, one would not expect to see another vehicle for hundreds of miles. We praised God for sending help. All of us together were able to push the station wagon and set it upright. As it came down, the windshield shattered in one spot on the driver's side. Was it even worth trying to save?

The proprietors of Fireside Lodge were gracious and let us have a room even though the Lodge was closed. They invited us into the restaurant. There we warmed up with coffee and hot chocolate. The restaurant was open even though the Lodge itself was closed. It was a rest-stop for truckers. It was too late in the day to bother anybody about helping us get our vehicle out of the snow. It was well past dark, and the temperature was dropping. We would retire to our room, kneel at our bed, and give thanks to God for his Grace.

The next morning, New Year's Day nineteen-seventy-eight, we got dressed and went down to the restaurant for breakfast. Perhaps it was the fact that we were not sitting huddled in our vehicle in the boonies waiting to freeze to death, maybe it was the fact of near forty below temperatures outside, or it might have been that because God had allowed us to have another breakfast, it was the best breakfast ever. The coffee seemed to say with each swallow, "be thankful for me." We were.

We noticed that two truckers were finishing their breakfast, so we went over and asked them if they could pull our Toyota out onto the road. They turned down our request because their insurance would not allow it. Dread and fear returned as well as a measure of disappointment. We had staked our hopes on some benevolent trucker but found none. Then along came Charlie. I do not know his last name, but—if he is still alive, then—, Charlie, we still—and always will—remember you. God bless you, Charlie.

He sauntered over to our table with an ear to ear grin and stuck out his hand to me. "Good morning. I'm Charlie. I overheard your conversation and I will pull you out." He followed us to the location of our "sled" and we observed that he had such a vast collection of chains. There were little chains, medium chains, big chains, and huge chains. He had done this before we surmised. He pulled one of his longer chains out, hooked us up, and out we came onto the road.

I reached for my wallet. "How much do I owe you, Charlie?" We would have been happy to pay him a lot.

"Just let me take your picture," Charlie replied. "I keep photos of all the folks I have pulled out of similar situations. It's my rogue's gallery."

"Ah ha! I knew it. I just knew it. We saw all those chains and guessed that you have done this before."

He chuckled and, after snapping our picture, he went on his way.

"We love you, Charlie."

We had brought a supply of Sterno, so I lit one and placed it under the oil pan of the Toyota. A few minutes later, I got inside and turned the ignition, not really expecting anything to happen. It started! Man, oh man, that engine sounded wonderful. We swapped vehicles. Rheta and Carrie drove the Ford. Steede and I took the Toyota. I drove looking slightly to the right where the glass was not broken. We were on the road again.

As we drove through Watson Lake at mile six-thirty-five, I became even more aware of how blessed we really were. The town's population was a bit more than fifteen-hundred people. It was the only town within eighty miles of the wreck site that had a doctor. Rheta got only a small scratch on her leg. Carrie was unscathed. It could have been so much worse. We tried to put the experience out of mind as we journeyed on. However, now we no longer were excited about such a beautiful experience. The sights were no longer as lovely as they once were. Well, they were, but now our focus was different. We just wanted this trip to end. We wanted to be in Anchorage. I wondered if my Jodie regretted having married me.

We married in nineteen-sixty-one and I moved us around a lot in my job. Promotions usually came with assignment changes. She followed me from our home in Fort Worth to Puerto Rico. She followed me from there to Key West and then to Atlanta. She was still with me when we moved to Oklahoma City. Now she was with me, at the risk of death, on this trip to Anchorage. I remember wondering, "Are promotions worth all this?" If she wondered too, she never let me know. She never complained. Once I applied to the World Meteorological Organization for a position in Katmandu, Nepal. Later I came to my senses and withdrew my application. She never said, "I don't want to go there." Oh yes, her love for me was proven long ago. But, I had never ever doubted that she did.

We still had some twelve-hundred-miles or so to go before reaching Anchorage and mostly it would be mountains, rugged ones, until we reached Customs at Beaver Creek. We pressed on. Teslin Lake is a seventy-five-mile-long lake, as straight as an arrow. It extends from northern British Columbia into southern Yukon Territory. Would we ever get out of these scary mountains? No, not for a long time.

Highway one would run along the bank of Teslin Lake for too many miles. The road had no guard rails and on one side was a lake and the other side sheer cliffs. Hadn't we been here before? Maybe, except for the lake. I sure do not want to run off into that thing. Oh, but wait, it was completely frozen over. I breathed a sigh of relief.

I do not remember many details of the rest of the way through the Yukon Territory. Stress levels had replaced excitement levels. Crossing into Alaska at Beaver's Creek gave us a welcomed and well-deserved respite from the mountains. Away from the treacheries of mountain driving, we now began to be enthralled by wildlife sightings and the breathtaking vastness of our largest state. Tok was coming up. Let's stop there.

Tok, Alaska is a small town just seventy-five miles into the nation's largest state on highway two. When we were there, Tok's population was around a thousand people. Highway two out of Tok continues to Fairbanks while Anchorage bound folks turn off onto Highway One, the Glenn Highway. We stopped there for coffee and a restroom break. I do not remember the exact temperature, but I think it was around thirty-five below. We were anxious to get to Anchorage, so we did not linger long in Tok.

Highway One skirts several mountain ranges and takes the traveler through a vast expanse of tundra. Tundra exists largely due to cold temperatures and short growing seasons. One might find a smattering of very short black spruce but, for the most part, there are no trees growing in tundra. There are short shrubs and bushes. In the distance of a little over three hundred miles, we saw an abundance of wildlife.

During the winter in Anchorage, the sun rises around eleven and sets around two o'clock. And, even then, it barely manages to get itself up over the thirteen-thousand-foot Chugach Mountain range. Of course, when I speak of the sun struggling to get over a mountain, it is purely from my perspective. The sun's trajectory across the sky was one of the first things I noticed. In Texas, the sun rises more directly in the east (depending on the season) and sets more directly in the west. In Anchorage in winter, it rises in the northeast and sets in the northwest, making a shallow arc across the sky.

Another interesting thing about Anchorage, or, for that matter, any cold climate city, is the fact that most houses have installed humidifiers to maintain comfort in the house. You know how it is. It is almost impossible to get warm when the air in the house is dry. Cold air is dry relative to warm air but pump a little moisture (a pot of water boiling on the stove) into the air and suddenly you find yourself lowering the heater thermostat.

The result of having a humidifier in the house when outside air is sub-zero is dense fog coming from each chimney throughout the city. When we finally got close to Anchorage, we could hardly see to drive. The fog was so thick I slowed down out of fear of caving in my front bumper. I do not know how we did it but we found our way to our hotel and settled in for the night. We were awakened to the sound of big snow plows clearing the streets. Seventeen-inches of snow had fallen our first night there. We arranged to house sit for a former National Weather Service meteorologist while he and his wife were in Hawaii. It gave us time to look for a house to buy. We found one and bought it. I still remember it well. Our new house was in the Flying Crown subdivision and was at twelve-eight-three-one Nora Street. The house was and still is a split level. You can google it on Google Earth and look at it. There is still the airstrip about fifty yards behind the houses, but we had no airplane, so it did us no good except as a selling point when we left.

I will not write about our trip back four years later since I think I already covered that experience in a post that I called "Christians Don't Gamble." I will say, though, if you ever decide to drive to Anchorage in late December or January or, for that matter, in winter, leave the dog and cat at home, install block heaters on your vehicle, take a survival kit, take chains, take a supply of food and water to eat if you get stuck or slide off the road (of course you will not need any more food if you go off a cliff), take blankets and extra socks, and…well, you get the idea. Do not.

Herman's Worthless Dogs

Whose cabin it was escapes me now. I believe it belonged to my brother A.C. It may have belonged to Henry. I just do not remember. What I do remember though, is the cabin itself. It, like most mountain cabins, was one room and had a fireplace, cook stove, and bunk beds. Although one individual owned it, back then nobody posted their land with keep out signs and cabins were left unlocked. Folks could be trusted to leave the place in good condition if they stayed there in the absence of its owner. We all figured it was ours too. We cared for it as though we owned it. We replenished wood for the fireplace and made sure the floors were swept, the beds were made, and the dishes were washed. Yes, siree Bob, it was our chalet.

The cabin was situated on the bank of Little Fourche Maline Creek which never ran dry even in the hottest summer. Deer and small game were plentiful within walking distance and the creek was handy for cleaning game as well as providing drinking water and water to cook with. The creek bed was rocky, so the water was naturally potable.

A daily routine for hunters was something along these lines. The cook, whomever was designated, rose early before the others were awake. He would stoke the fire and add more wood when needed. The cook-stove was fueled by propane, so the cook would light the oven to pre-heat it for biscuits. The next step in his routine would be to pull out a skillet and began cooking the breakfast meat, usually bacon. The smell of bacon frying and hot coffee was always enough to roust even the laziest sleepyhead among us. A typical breakfast consisted of bacon and eggs with biscuits and gravy. And, of course, hot coffee was an essential part of breakfast.

By the time breakfast was ready, everyone was up and chomping at the bit to "go out and get em." While the cook put the finishing touches to breakfast, the others were planning their day. I think I will go hunt Yancy today. Where are you guys going? Each would name their hunting spot. There was a lot of laughing and joking as we filled our stomachs. After everyone had eaten, someone cleaned up the dishes and the others headed out to hunt.

Herman was one of the regulars and always brought his horse in case someone got lost in the mountains. His old horse rode double so, that way, he could offer the weary hunter a ride if he was lost. It was a sight for sore eyes for more than one missing hunter. But Herman's horse served a different purpose on one day. I have always been a practical joker and that ole sorrel was just what the doctor ordered for a prank that would evoke laughs for years to come.

Herman always hunted with dogs. A good deer dog would strike a trail and then run the animal in a circle, bringing it by its master. Herman's dogs were good. Very good. So good, in fact, that during every trip we all had to endure his frequent brags about their value. Those dogs could do no wrong. Even the Queen's fox hounds could not hold a candle to Herman's. That mindset of Herman's was a perfect prank in the making the next morning.

The day before my prank, someone had killed a small deer and butchered it for camp meat. That sort of thing was a common practice in the nineteen-fifties. Every morning found Herman out, as soon as daylight made its appearance, leading his horse to the creek for water. It was a routine, one he never failed to follow. My prank would not fail.

The evening before, after lights went out in the cabin on Little Fourche, a diabolical plan began to unfold. To relieve one's self it was necessary to go outside and use the outhouse about thirty yards away. I waited until I heard snoring and then got out of bed, put on my clothes and shoes (to avoid snake bite and mosquitoes), grabbed a flashlight and went outside. Nobody would have suspected that I was doing anything other going to the outhouse.

I found where I had stashed the four legs of the deer that fed us the day before. I tried to imagine myself as a curious deer wanting to know what the humans were doing in there. Then, very carefully, I positioned the feet in the fresh mud around the cabin. It had rained the night before. I was convinced that God was helping me with my prank. He would laugh too. Oh wait! The Psalmist said that God looks at the plans of men and laughs. But, the context makes it clear that bad men are spoken of there. I was not bad, nor was I a man. I was just a skinny kid who loved to play pranks. God would forgive me.

I began planting the tracks down by the creek at the "watering hole." Herman was a hunter, a good one. He would notice the sign immediately. If I could have found deer droppings I would have put a pile right there. After finishing my handy work, I shined the flashlight on the ground all around. "Now that there is plum good work," I whispered to myself. I went back to bed and begged to morning to come early. It may have been my best caper every. I do not think I got much sleep that night. While awake, I chuckled to myself with each thought of it and how Herman would react. Even when asleep, I must have chuckled. Such an invention required it.

The sounds I heard the next morning were all too familiar. Herman was up. The horse was in a small makeshift lot next to the window where Herman could keep an eye on it. His incomparable dogs were in their kennels. A terrible thought occurred to me. Would my elaborate plan fail yet? Oh, please no. I worked too hard perfecting it. It was not good light yet. Would Herman even see the tracks? I had planted them deep to reflect the weight of the deer.

Alas, my worries had been for naught. Herman had piddled around enough so that good light would easily enhance my masterful trick. The cabin door opened and the cold outside air rushed in. It quickly closed, and Herman walked around the side of the building to the horse lot. He snapped a lead rope into the horse's halter and off they went to the creek. I had expected an instant outcry, but none came. How could he have missed the tracks? Herman loved to hunt squirrels so perhaps he was looking up at the trees. At any rate, he would spot them. He was too good a hunter not to see them.

Still wrapped up in my covers, I waited. Then I heard the commotion that my sense of humor would feed off for years to come. You see? This happened in the nineteen-fifties and I am still laughing as I tell about it. The noise I heard was that of a startled horse jumping around and dogs barking as they were verbally set upon by their master. Herman was never cruel to animals, but he must have come as close as could be without shooting them that day (which reminds me of the two guys who jointly owned a coon hound that would not hunt. One man said to the other, "I think I'll shoot my half)." The problem was, Herman did not co-own the dogs. They were his and he had been proud of them beyond description. I wondered, would he still be?

As the hounds slunk back into their kennels, the ranting subsided. Dang! I expected more. Then, a moment later the door to the cabin burst open and the ranting began again. If anyone was still asleep before, they sure were not now. Someone asked him what in the world was wrong. It sure was not me who had asked though. I was sure that I was soon to be murdered so I pulled the covers up over my head and waited for the Grim Reaper.

Herman stomped and yelled and described the value of the dogs in a way that was exactly opposite of his earlier assessment. "Those hounds couldn't find a rabbit much less a deer. They had to be totally worthless. Had he been feeding those mangy four-legged creatures?"

I'm sure he must have been thinking, "Should I shoot them, give them away, or just turn them loose to fend for themselves?"

Herman opined that the least he could say about them was that they were worthless. Then, I guess he must have remembered all the kind words he had spoken about them, because his next words were, "And I thought they were supposed to be the best dogs around." Disappointment weighed heavy on his heart and his tongue. I felt sorry. I almost regretted having done such a thing.

"Wait until I see the guy that sold them to me," he growled.

Uh oh, this was going beyond where I meant to go. I didn't want him to kill someone because of my trick.

I quickly repented. I got out of bed and showed Herman the four deer shanks I had used to plant the tracks. "I did it, Mr. Suttmiller. I played a trick on you." He stared right through me. I just knew he was thinking to himself, how can I kill this kid and get by with it? But, we were friends. The old man forgave the young boy. Everyone got a big laugh. Herman laughed too. All was well again in the little cabin on Little Fourche Maline.

Henry's Headless Turtle

When I was attending the University of Texas at Arlington, I was required to have eleven semester hours of German, French, or Russian to graduate with a BS degree in mathematics. I chose German. We had to have eight hours of German grammar and three hours of German literature. In the literature course, we had to read and analyze short stories and write a report. In the report, we were not allowed to resort to the use of English. That was a real trip. I read several short stories by the Brothers Grimm but none of them were about a headless horseman even though there are two such stories by them. In American folklore, there is also a story about a headless horseman. Everyone remembers "The Legend of Sleepy Hollow" by Washington Irving.

I somehow doubt that Henry had ever read either the Brothers Grimm or Washington Irving, but he had a tale that entertained me almost as much. I say it entertained me because I was just a young boy who was easily entertained by Mr. Wilson. And, I had not read the tale of Sleepy Hollow. So, naturally, when I did read it, I thought of Henry and his tale about a headless turtle.

It was on one of our three-day fishing trips that I first heard it and I suspect that not one such trip was taken after that but what I wanted a repeat performance by Mr. Wilson. Henry loved to tell stories and he was a master storyteller. It was not a very long story, but it had an ending that always made me laugh.

Henry took his two boys with him on a fishing trip when

they were small. They set up his tent as I had seen him do many times. After staking down the tent, Henry put a bait on each boy's fishing pole and sent them down to a small neck in the creek where the water was not too deep and where he could keep an eye on them. He, being a more serious fisherman, dropped a line in his favorite fishing hole.

After some time, the boys got bored with fishing and started exploring. Henry thought to himself, "well if we are going to have fish for dinner tonight, I guess I'm going to have to catch them." He continued to cast an occasional eye on the boys as he tried to coax a fish onto his line.

It was not long before the two youngsters came running up to him as excited as if they had been given a new bicycle for Christmas. "Daddy, Daddy," they yelled out as they ran. Henry figured that they must have seen a sixty-pound catfish by the way they were carrying on. But they had not. It was much better than that. What the two boys had spotted was a turtle with no head.

As excited as all get-out. those two scamps insisted that their daddy go with them down to the place where the boys had spotted the headless turtle. Figuring to appease them, Henry walked with them to the site.

"Daddy, Daddy, see it?" one of the boys exclaimed as he pointed to a large snapping turtle.

Only problem was, that danged turtle didn't' have a head. No siree Bob, its head had been plum cut off, leaving only a loose flap of skin that had apparently healed over. Well, that sight got Henry's attention.

He and the boys stood there and discussed how that could have happened. How could that ole turtle live without a head? Henry surmised that perhaps insects and small frogs felt sorry for it and crawled up inside the turtle. After a brief period of being amused and bumfuzzled, Henry turned and started walking back to the fishing pole that he had set in the bank.

"Hey Daddy, aren't you going to kill it?" questioned one of the boys.

"Yeah, Daddy, you must kill it," the other lad chimed in.

Henry's reply? "Hell, boys, I don't know how you're going to start," he stated, as he went back to fishing.

Wash Day in Rural Oklahoma in the Nineteen-Forties

"Dear, does this go with the coloreds or whites?" In rural Oklahoma in the nineteen-forties such a question was never heard on wash day at the Mankin house. Dad never helped with washing and Momma knew where each clothing item went. My older sisters helped.

A typical wash day at our house back then found us all down by the spring. Dad had discovered an Artesian flow of water in one spot and turned it into a well by digging down seven or eight feet and walling the sides with rock. We called it "the spring." It had a strong flow of water and never went dry. Beside the well was a galvanized gallon bucket with a rope tied to it. A bit of practice was required to drop the empty bucket in such a manner as to fill it with water. The bucket naturally wanted to turn upright as soon as it hit the water. But, after a few tries, anybody could become skilled at filling the water bucket and hoisting it to the top. We carried all our water needs from the spring.

But Momma and the older girls did the family wash at the spring because it was about a hundred yards from the house so washing clothes there saved lots of time and energy and it was also a beautiful place to work. Two massive sycamore trees stood always ready to provide refuge from the hot sun on summer days. As a small boy, I always thought that those large sycamores created the wind. I had seen folks fanning themselves with cardboard fans on hot days. I equated the moving of a fan with the movement of leaves on those big trees. Naturally, my hypothesis was, "the trees created the wind." I suppose that was my first failed prediction. There would be other wrong forecasts years later before my career as a meteorologist ended.

I would discover early in my studies of physics that a difference in air density was the cause of wind. Because the spring was located at the base of a ridge which I have often mentioned, a breeze usually began around mid-morning when the slopes heated up and air began to chimney up the hill, thereby drawing air through the sycamore trees. Many years later, as a fire-weather meteorologist, I would see upslope, downslope, up mountain, and down mountain wind patterns all caused by the sun.

A huge three-legged cast iron kettle sat off to the side on a level spot of ground. Bricks were used to raise the kettle up high enough to build a fire under it. Momma and the girls would carry two number three galvanized washtubs to the wash site and as soon as the water was hot to her liking, Momma would transfer it to one of the tubs. The other tub, she filled with cold spring water. She always brought a bar of lye soap (lye soap making is a story for another day) and a rub-board. The cleaning process was quite simple. There were no gears to freeze up or break, no fuses to blow, nor coins to put into a slot. There was only the continuous back and forth rubbing of the soapy clothing on the old rub-board powered by muscles.

Once the garment was clean, she would transfer it into the tub of cold rinse water that contained "bluing," an agent to brighten the white clothes. After sloshing the garment in the rinse water, my older sisters wrung the water from it by twisting it as tight as possible. After washing, rinsing, and wringing the garment, it was hung to dry on one of two or three clothes lines stretched between trees. The sun dried the wash quickly and acted as a bleaching agent for white things. Wash day was a tiring day for Momma and my older sisters. For me and my youngest sister though, wash day was fun. We got to play in the cool shade of the sycamores and drink cool spring water. Our day of assigned labor would come eventually. For the moment, however, we were "foot loose and fancy free." Wash day was fun. We looked forward to the next such day.

El Comandante

In early nineteen-seventy, I received a letter from the director of the National Weather Service Southern Region Headquarters. "Dear Don," the letter began. I have selected you to fill the Hurricane Preparedness Meteorologist position in San Juan, Puerto Rico." We were so excited. We had found it hard to wait until the vacancy closed. Then the personnel committee had to evaluate all applicants and come up with a list of the ones they deemed to be highly qualified. Once the list was submitted to the director, he would select the applicant he felt was best. Little did we know at the time, but God was at work arranging events and ordering our lives.

Puerto Rico was a popular assignment and I remember thinking as I wrote up my bid, "Why am I even bothering to fill out these papers?" But I did, and it paid off. I got the job. We immediately contacted a real estate agent and listed the house and forty acres for sale. My date to report for duty in Puerto Rico gave us little time to sell our place. We were expecting to have a financial struggle until we could find a buyer for the property. It would be hard to make the farm payments and rent payments in Puerto Rico. But God had other plans. He sent Bobby and Mary Fulks soon after we listed our place for sale. They bought forty acres and the house. The proceeds were enough to pay off the remaining acreage. We headed for Puerto Rico debt free. Steede was almost six and Carrie was eleven months.

I will never forget my first breath of tropical air. It was April and nights were cool in Springtown, Texas. Not in San

Juan though. The air was heavy. I wondered, "have we made a mistake? This humidity is terrible. Will we be able to adjust?" We would indeed but at the time, I doubted it.

We got into one of the many taxis that were lined up outside and headed for our guest house. We were pleased with the arrangements that Dr. Jose Colon, Meteorologist in Charge of the San Juan Forecast Office, had made for us. The house was built to take full advantage of the sea breeze by day and land breeze by night. So, with an almost continuous brisk breeze blowing, the humidity that I had dreaded seemed innocuous.

A restaurant called Cecelia's was only a few yards from our guest house. Dr. Colon said it was good. We tried it. Indeed, his recommendations were spot on. There we tasted arroz con pollo (chicken with rice) for the first time. And, it was at Cecelia's that we had our first mofongo (a delicious concoction made of plantains that have been fried, mashed with garlic, shaped into a ball, and served in a pilon (the mortar part of a pestle and mortar). The dish is fantastic. Asopao and paella were newly discovered favorite Puerto Rican dishes.

Outside coffee bars were plentiful. One was located near my office. I had my first café con leche there. A mountain range, the Cordillera Central runs west to east through the middle of the island. Some of the best coffee in the world is grown there. It would be almost a year before Starbucks would arrive on the scene. What we now call a latte was perfected in Puerto Rico as far as I was concerned. There was no foam on a café con leche. Instead, a good strong brew was poured into a cup, about half full, and then hot milk was added. Very soon, a skim was formed on the top of the coffee. Perhaps it was the strength of the coffee or maybe it was because I was drinking the Joe on an exotic island where the beans were grown, but I sure fell in love with café con leche.

The guest house very quickly became too cramped for the comfort of all so we went house hunting. The National Weather

Service required me to take a crash course in Spanish at the Benedict School of Language. Thinking it to be a great idea, Rheta enrolled also. We figured it would be best for us to move out into the island away from the cluster of "continentals" as we were called. We would learn the language faster that way. We rented a three-bedroom house in Rio Piedras, some twenty minutes from the airport where I worked. We almost did not need an alarm clock there. Early each morning, a street vender would push his cart down our street, Calle Cipres, calling out, "huevos, pollos, pan, pan de agua" (eggs, chickens, water bread). After we had lived there a few days and our children had made friends with neighbor kids, our son called out in reply to the street vendor, "huevos, pollos, perros, y gatos (eggs, chickens, dogs, and cats). We laughed.

El Comandante race track was only a mile from our house in Rio Piedras. Being the lover of horses that I am, we went to the races just to watch the magnificent animals run. I started noticing that a certain number assigned to a horse in a certain race seemed to win most often. Could it be that I had hit on something? Fresh out of the University of Texas at Arlington with a mathematics degree, I decided to investigate. I gathered up about three hundred old racing forms from newspapers and other sources and went to work. I worked up a Poisson distribution (teachers, among others, know this as the bell curve) and was anxious to try it out at El Comandante.

One Sunday afternoon, I was sitting at the kitchen table with the latest racing form. Lo and behold! My method was predicting a long shot to win. Jockeys in Puerto Rico were reputed to be inclined toward fixing races, a "you can win today and I will win on such and such day" sort of arrangement. In fact, I had seen on many occasions what I suspected was a jockey holding his horse back as he neared the finish line. "That's it! That is the reason for the clustering of certain numbers in certain races. Number five, you win in the third race today (5/3) and number eight, you can win in the fifth race (8/5)." But would it work?

"Rheta, my method shows number five horse will win in the third race and it is almost time for the race to start. Let's go," I urged. We hurried to the tracks. The race was about to start. I would not get to the ticket gate in time if I parked the car, so I asked Rheta to park it. I got to the ticket sales cage just in time to put two dollars to win on horse number five in race three. Remember, number five was a long shot, a very long shot, something like eleventh in a field of thirteen. I left the ticket gate to go find Rheta. An overhead monitor revealed that the race had started. The horses were running. But where was my horse? I was looking for him in a bunched-up group near the front. Oh wait! Dang! He is running dead last. "Well so much for Simeon Denis Poisson," the French mathematician who devised the Poisson distribution. I walked to the bleachers looking for Rheta. Ah, there she is. I glanced again at the monitor. It was flashing number five. My horse had won. I walked away fifty-six dollars richer.

No clear-cut winner was suggested in the next four races, so we headed for our car. When we got to the car, we saw a guy standing by his Cadillac urinating in the parking lot. For crying out loud! We never did things like that in Oklahoma or Texas. I understand that El Comandante has lost most of the glory it once had in its heyday. When races do happen, it is usually a very small field. Well, El Comandante, you can't win them all, as I found out the very next time that I bet on the horses. Sad. Even sadder when I abandoned Monsieur Poisson.

Always Be Gentle

"A gentle answer turns away wrath…"
~ Proverbs 15:1

Years ago, when we were living in Springtown, we went to dinner with our dear friends Joe and Judith Osburn. We discussed where we should go for dinner and decided on Mark Dean's Bar B Q. Mark Dean started selling his wares from a small elongated building with a corrugated roof and a front that was screened at the top and had wooden hinged window covers that could be raised to let in fresh air and lowered to keep out inclement weather. The food he served was so good that his business flourished. I do not remember how long it was but eventually demand for Mark Dean's Bar B Q outgrew his little store where West Main branched off from Jacksboro Highway. So, it moved to a new location in Azle. Our first visit there was a hoot.

Rheta drove a Cadillac at that time and we went to dinner in her car. I drove. We pulled into the parking lot at the restaurant, went in and enjoyed a rib dinner, and returned to our car. I had trouble finding an exit from the parking lot. I made several rounds looking for a way out? Dang! Did they entice us into this lot to trap us? Where was that silly exit? As I slowly drove around looking for the exit, a man in a beat-up old car stopped beside us. I only mention the condition of the car to suggest a possible motive for what was about to transpire. We were pointed opposite to one another, so the man rolled down his window and gave out with a few expletives followed by "you don't know what the hell you are doing, do you?" I smiled at him and replied, "well how in the world are you?" Completely disarmed, he responded. "oh hell, you damned Cadillac drivers are all alike" and drove off. For crying out loud friend, you could at least have pointed me to the exit.

If You Ain't a Backee Chewer, Leave the Beechnut Alone

When I was in my teens, our place in Panola, Oklahoma had a large hickory tree in the back yard. One afternoon, my sister Nadine, threw a Hickory nut at me. Her aim was impeccable. Wham! One moment I had two maxillary central incisors and the next moment I had one and a half. Her perfect aim took out half of one. Dad took me to the dentist in Wilburton. He capped the tooth. Dentistry sixty years ago was vastly inferior to that of today. The tooth abscessed and so did the one next to it. Both teeth had to be pulled. All I wanted that Christmas was my two front teeth (from a song titled 'All I want for Christmas is my two front teeth' by Donald Yetter Gardner). The dentist constructed a partial plate.

The partial plate lasted through high school as well as through Air Force boot camp and weather observer school in Air Weather Service. I was stationed at Carswell Air Force Base in Fort Worth. One of our responsibilities was to take weather observations which consisted of measuring the height of cloud bases, dry bulb and wet bulb temperatures from which dew point temperature and relative humidity were calculated, sea level pressure, altimeter settings, and wind direction and speed. Another very important part of our duty was that of plotting weather data on surface and upper air maps. We also plotted data on Skew T Log P pseudo adiabatic diagrams.

To measure the base of low clouds during the day we used

small balloons that had an ascension rate of twelve-feet per second. At night, we used a clinometer which we called "the beer bottle" since it resembled a bottle. A small concrete pad marked the spot where we would stand to measure cloud bases at night. One thousand feet from the concrete pad, a spotlight shown on the base of the low clouds. We would look through the "neck end" of the clinometer and fix it on the spot on the cloud base. Then, we tightened a sliding arm that pointed to angle values inscribed on the scale part of the clinometer. Knowing the distance from the concrete pad to the spotlight, one thousand feet, we would go back into the station and look up the tangent of the angle we measured. The tangent of that angle multiplied by one thousand feet would give the height of the cloud bases.

One dark night, I was working with a kid from Abilene. He chewed Beechnut chewing tobacco. We were plotting Skew T diagrams. Jerry pulled out a new package of Beechnut and offered some to me. I took a wad of it and popped it in my mouth. We continued to plot diagrams. The clock on the wall said it was time for an observation of the cloud height. It was my turn with the "beer bottle." I picked up the clinometer and headed out into the dark night. I did not take a flashlight. Just before reaching the spot where the measurement was made, there was a rock wall about three feet high. In the dark, I got to the wall before I expected. I fell off onto the ground below. When I picked myself up, there was not much tobacco left in my mouth. It was suddenly in my stomach.

I walked back to the station and calculated the cloud height and went back to plotting maps. I did not plot for long, however. Moments later I was hanging my head over the toilet bowl heaving my guts out. I retched and retched. Then I retched again. I wanted to die.

The agony finally over, I lifted myself to my feet and felt an empty space in my mouth. I had upchucked my partial plate down the toilet. Oh well, I never liked that partial plate anyway,

And, besides, the Air Force made a bridge that lasted me until I had it removed about ten years ago when I was teaching students with behavior problems. It had gold edging, very dated. Now perhaps my student would stop calling me "Old Gold Tooth."

I would have done anything to avoid sleeping in a dirty bed.

Before the days of hurricane evacuation maps, Dr. Colon, Meteorologist in Charge of the San Juan office assigned me to make a survey of exposed populated coastal areas in Puerto Rico and the U.S. Virgin Islands. I used U.S. Geological quadrangle maps and the Saffir-Simpson disaster potential scale to determine how far inland storm surges would penetrate during hurricanes of various strengths. I tried to maximize my time by scheduling speeches on hurricane preparedness in the various towns I would visit during my survey. I had given a hurricane preparedness talk to a civic organization in Ponce, Puerto Rico across the island from my home in San Juan. Back in the seventies, there was no four-lane expressway across the Cordillera Central as there is today.

When I think back to that time in my career, I shudder to think of what could have happened to me. I left Ponce around nine o'clock in the evening and headed for San Juan. I was alone and the night was pitch black. The distance from Ponce to San Juan today via the expressway is about seventy miles. It takes about an hour and twenty minutes to drive the distance now but back then it took about three and a half hours. The narrow two-lane highway caused the bravest of drivers to slow to around twenty to thirty miles an hour. Steep drop-offs and few guard rails were common along the edge of the highway. Steep slopes, hairpin curves, and dangerous switch backs made me wish I had not embarked upon the night trip to San Juan.

Why didn't I get a motel and stay the night? I had planned to do just that. Before giving my speech to the Lion's Club, I made a reservation at a motel in Ponce. My meeting was a dinner meeting and it was around nine-thirty when I got to the motel. I went to my room and turned back the sheets anticipating a good night sleep. What I saw immediately disrupted my plans. Between the sheets, I saw several curly hairs. I can tell you for sure they were not toe hairs. I was glad I had already eaten dinner. I promptly complained to the manager and left. The drive was stressful, but I arrived safely home and gave thanks to God. Would I have driven it again? You bet I would have. I would have done anything to avoid sleeping in a dirty bed.

Madchen, Madchen, Madchen

She ranked high on our list of dogs we liked. We bought her after losing Kahn. Madchen, "little girl" in German, was a registered Rottweiler. As we learned with Dobermans, the breed gets an unfair label of being a mean dog. She was anything but mean.

We lived in Springtown and we kept her in the house. Well we did until she brought fleas into our domain. Then she became an outdoor dog. I will never forget that experience. We wondered why every member of our family had itchy legs. Blame it on Madchen. She had made friends with a gazillion fleas and wanted them to live with her. We set off several exterminating bombs and got rid of those little vermin.

Madchen loved people. She was convinced that everyone loved her and wanted to play. Her favorite game was to get your arm and drag you around wherever she thought you would like to go. She was massive enough to crush your arm but gentle enough not to.

Gentleness was the thing most remarkable about her. Only our friends knew that though. Strangers would not get out of their car. She loved playing with our cats and they loved Madchen. Another game she loved was to run at a cat and then, when it took off running, run alongside of it and eventually run past and stop in the path of the feline. She would have made a good cutting dog. In the winter time the cats would curl up against her to stay warm. Our Buff Orpington rooster did the same. He also laid next to her throughout the year. When Madchen got too old to hear, a coyote snuck close and grabbed the chicken before she could defend it.

Our son's youngest boy loved Madchen. He was just learning to say words and he would say them three times. When our son brought the family out on weekends, Allen knew he was close to Madchen's house when they got to Midway Road. He would announce "Madchen, Madchen, Madchen. Once when they came for the weekend, Rheta carried Allen out to see his beloved dog Madchen who was laying by the barn door. As usual, he announced, Madchen, Madchen, Madchen" but this time the big Rottweiler did not respond. She just laid there. Rheta hurried away to distract Allen. Madchen would never greet him again. Nor, would she chase another cat or lead me around by the hand. We buried her in the shade of the large live oak tree where we had laid Kahn to rest. Now when I listen carefully, I think I can still hear baby Allen call out "Madchen, Madchen, Madchen."

Stay Between the Lines, But Which Lines?

In nineteen-fifty-seven when I first came to Texas, even an old worn out road map would get you where you wanted to go. The only time you needed to get another one was when you marked the map up too much or lost it. Roads just didn't change all that much. As for cities, well, Tulsa, Ft. Smith and Oklahoma City were the most populated places I had ever been. Navigating highways from one city to the next was duck soup.

And, for the most part, cities were safe. At least they were in Texas. Folks never thought of locking their doors at night. You could even safely sleep in your car. I know, because I did. I arrived in Fort Worth in the evening of the day before I would report for duty at Carswell Air Force Base, a Strategic Air Command base. At one hundred and twenty dollars a month salary, a hotel room was out of the question. So, I slept in my car.

I liked Fort Worth. I quickly settled in to my new life in Cowtown USA. I recall one day that a fellow airman and I drove highway one eighty-three. We were going to check out Dallas. Now there was a city for you. No place for a country boy. The time of year was nearing the Christmas season. When we got to loop twelve, I blasted right through a red light. I guess I thought it was part of the Christmas decorations. At any rate, I looked over at my buddy and he was as white as a sheet. Wilburton, Oklahoma on Saturday night was no match for Dallas. Wilburton only had one signal light.

I could have run every red light in Dallas and suffered less stress than I did this morning though. I drove to Irving to pick up a part for my Volkswagen Tiguan. The place is just off highway one eighty-three on Story road. That section of one eighty-three has always been under construction of one sort or other. The soil is very unstable and frequently the road, with time, buckles and heaves. At times, you must roll down the window and stick your arm out to convince yourself those are not frost heaves like in Alaska and Canada.

Things went well until I turned off three sixty onto one eighty-three. Then, the realization that I was in the middle of a stream of big sports utility vehicles and eighteen-wheelers surpassing the speed limit by far too many miles per hour struck fear in my soul. A town full of zombies chasing me could not have caused me more fright. That part of the highway has been repaired numerous times and each time, it seems, old lane stripes are left as new ones are added. They crisscross and merge in such a way that makes it hard to stay between the lines. If you do not exercise great care, what you think is your lane will direct you right into a line occupied by a monster eighteen-wheeler, certainly no place for a Volkswagen Tiguan.

After being tail-gated by speeding big dogs and trying to share a lane with truck drivers bent on killing me, I finally pulled into my driveway a complete blithering idiot. And, by the way, John Denver you rascal you. Why can't I get your song out of my mind? Is it forever and inextricably locked into my brain? Oh well, I like the song. So, "Take me home country roads."

Roasted Chicken Tasted Better in Seventy-One

It is interesting how ordinary food tastes different depending upon such things as mood, setting, familiarity, and such. Take for example the tiny U.S. Virgin Island of St. John. This part of God's creation never gets old to us. We have always enjoyed lunching and dining at places like The Quiet Mon, Woody's and Morgan's Mango, all in the little town of Cruz Bay. Other places we enjoy are Miss Lucy's across the mountains and Skinny Legs which overlooks a small bay filled with sailing boats.

My first time on St. John was without Rheta. I had to give a talk and present work I had done to facilitate evacuations during hurricanes. I flew from San Juan, Puerto Rico where I was based, to St. Thomas where I met the director of civil defense for the U.S. Virgin Islands. He had his private government boat waiting for us. We boarded and made our way to St. John where our arrival at sunset found us both hungry. He knew just the place where we would get a bite to eat. I trudged along with him through sand and coconut trees to a little wooden shack. The screen door was barely on its hinges as we entered at the beckoning of a little black lady whom he called "Momma."

Leon asked her whether she had anything to eat, seeing that she was obviously closing for the day. With a laugh that exposed her beautiful white teeth and warm heart, she responded that she could always find something for Leon. She made me feel welcome too. She seated us at one of the small square tables covered with oil cloth and started plating some food for us. I do not remember what she served with it, but it was the best tasting chicken I had eaten except of course for Rheta's baked chicken.

I think perhaps that "'Momma's" roasted chicken served as a magnet to bring the Mankin family back to the islands numerous times in the succeeding years. Each time there I have always thought of that first experience on St. John and I think my assessment is true. Roast chicken tasted better in seventy-one (except of course for Rheta's).

On My Way to China

I have this marvelous app for my droid. It is called Google Sky. You can point it at any star and it will tell you the name of the star or planet. Interestingly, if you point it down below the horizon it will identify the stars that have already set. After sunset, you can see where the sun is on the other side of the world and watch it as it approaches sunrise again. That got me thinking. When I was a very young boy I learned that China is on the other side of the world and, I was told, if you dig straight down you will eventually come out in China. You know, I do not know how many of my dad's shovels I wore out trying to go there. Someday I want to visit China, WITHOUT a shovel.

Time in a Box

In my mind, I can still retrace every step that I took when I was twelve and on my way to the spring. The well-worn path began only a short distance from the back porch of our house. I clearly see it even now. It is the same house I wrote about before. You remember, don't you? It is the one that sat at the foot of the ridge where I often laid awake at night listening to the sound of hounds running a fox, the one where I would sometimes lay awake and watch the faint, flickering flames of fires on the mountainside. It is the house where my dad would send me to the kitchen to get matches which he would use to start a fire burning in his Prince Albert-filled pipe. It is the same back yard where he would lope down the trail toward me from the barn, making horribly frightening sounds in the dim light of eventide as I gathered kindling wood that my mother would use to build a fire in the cook-stove the next morning.

I am retracing those steps now. From the back porch, I make it about fifty or sixty twelve-year-old steps to a cluster of elm trees where the trail then curves and straightens out on a gentle down-hill slope the rest of the way to the spring from which we carried water for home use and where my mother washed our clothes under giant sycamore trees. Something on the order of twenty more steps is where I stopped to bury the box. The trail sloped off to the right toward a small ravine (not the scary ravine where the trolls and hobgoblins stayed). It was the perfect spot for my box of treasures.

The box was constructed of tin and was perhaps four by six

inches. Admittedly, it was too small to hold many things but just right to accommodate my treasures. After all, I was only twelve. How much could a twelve-year-old boy accumulate in nineteen-forty-nine? In the box, I placed a piece of paper. On the paper, was my name and the date. I dug a hole, placed the box in it and covered it up. I was sure that one day, many years later, I would return to uncover the box and check on its contents, but I never did. I presume it is still there on the side of the trail that leads to the spring. Or, perhaps, as with all things eventually, time has taken its toll.

Another item that I placed in the tin treasure-chest was a penny. Pennies were hard to come by in those days, but I had one that day. Maybe it was part of my earnings from selling garden seeds. I do not remember the date the penny was minted but it was like the one that my mother gave me every Sunday, when I was very small, to drop in the offering plate when it was passed around in church.

No doubt the most important thing that I hid away that day was a flat rock. Oh, it was no ordinary rock. It was nicely shaped and streaked with different colored minerals. I had used another rock with which to scratch two names, mine and the girl I was sure that one day I would marry when I grew up. Little did I know at the time, thirty-six miles away in McAlester, a cute little nine-year-old girl in pigtails played in her yard with no thought of a boy from Panola. If only I had known her then! Her name would have appeared with mine on that rock. Ah, but, although the rock may never be unearthed, her name is on a paper with mine and the rings on her finger and mine replace that rock. So, rust away little tin box and let the waters of time wash away those memories at will. I have plenty, much better ones that I have acquired over the past fifty-six and a half years with Rheta.

Angels Who Guard

It has been said that every person has a guardian angel. I do not know if it's true or not. All I know is, I have one. She used to ride with me. Now I ride with her. She sees better than I do. I have macular degeneration and only drive in places where I am familiar with the surroundings. So, she has become my chauffeur. And, a mighty good one she is indeed.

When I do drive, and she is with me, she keeps an eye out in case I do not see the car merging onto the roadway ahead of us. When we are walking on the sidewalks in a city that is unfamiliar to us, she frequently looks back to see if I am following. I often stop to take pictures and fall behind everyone else. In those instances, she stops and waits for me. She knows that I get lost in my back yard.

Recently, on the island of St. John, U.S. Virgin Islands, I walked down the road that led to our villa. The flowers along the road were astoundingly beautiful and I wanted to capture that beauty on my camera. One photo shoot followed another until I found myself on a different road. I did not recognize the surroundings. About fifteen minutes later, I saw the roof of Villa Pinache. I was on the right road and rather proud of myself. This time, my guardian angel did not have to rescue me.

"Did you get some good pics?" she asked as I entered the great room.

"Oh yes," I replied, "But I got a bit turned around."

She smiled. I read her thoughts. "Why am I not surprised?"

Glad to be safely back among family, I kissed her on the cheek.

She smiled again. Again, I read her thoughts. "He still needs his guardian angel."

Now That, Friend, Was A Real Hanging

We boarded the train in Paris and traveled to Caen where our personal guide met us and took us on a tour of Omaha Beach and Utah Beach in the Normandy region of France. It was in the little town of Sainte-Mère-Église that I met John M Steele. Well, I did not really meet the man, I only met his story, and a "whale of a tale" it is indeed. If my memory serves me, our guide told us that the town dates to the eleventh century, but the story of John Steele took place during World War II.

His story began shortly after midnight on June sixth nineteen-forty-four. Private Steele was a paratrooper in the five hundred fifth Infantry Regiment when General Dwight David Eisenhower ordered soldiers of the eighty second, one hundred and first, and five hundred fifth airborne divisions to mount an attack on the little town. Taking Sainte-Mère-Église was critical to defending against a German counterattack on allied troops landing on the beaches of Normandy.

A bombing run preceded the dropping of troops into the town of about twenty-five hundred. Many buildings were burning as a result. Sadly, the fires lit up the skies and provided "sitting duck targets" for the German soldiers below. Heavy Allied casualties resulted. One near-casualty was John M Steele.

As Private Steele swayed back and forth on his descent, his parachute got hung on one of the numerous spires on Sainte-Mère-Église, Church of Saint Mary. Steele was only one of many of our boys whose parachute got snagged on things like lamp posts and trees. Many became target practice for the German soldiers below

Very soon the town would be taken by forces that had landed on the beaches together with those who had landed and successfully cut loose from their parachutes. However, it would not be soon enough to save many of the paratroopers hanging from trees, light poles, and building spires. Once rescued, Steele would describe his plight.

He wondered, "Should I cut myself loose and face possible death from the fall or certain capture if I survive the fall, or should I pretend to be dead?" He chose the latter. For over two hours, Private Steele hung there until, finally, German soldiers spotted him and took him captive. They did not enjoy their victory long however, because Allied forces overtook the German forces and freed Steele. He returned many times to Sainte-Mère-Église in succeeding years before passing on May sixteenth, nineteen-sixty-nine. To that little town, John M Steele is still a hero and many references to his name can be found there. Restaurant Le Normandy has a statue of him hanging from the ceiling parachute and all.

The church still stands there today, having come through the war unscathed, for the most part. Hanging from one of the spires is a replica of John Steele and his snagged parachute. I stood there a couple of years ago and lost myself in the moment. My imagination ran wild. What was it like, for him? Was he paralyzed, with fear? Did he think of his family back in the states? Did he pray? But, before I could wonder more about John Steele's plight, I was summoned back to the others. As I walked away my last thought was, "now that, friend, was a real hanging."

Hello, Momma, I'm Gene Poag Jr, and I'm Here to Sing a Song for You

He grew up in Macon, Mississippi, this singer and song writer. I first met him in June of nineteen-seventy-two when I oversaw the Key West, Florida office of the National Weather Service. Hurricane Agnes was forecast to pass close to Key West. My office was short staffed, so I asked Headquarters for help. They sent me Gene Poag.

I do not often make instant friends but before the day was over, Gene and I would begin a life-long friendship. A few years later our paths would cross again in Alaska. It is not difficult to understand why we bonded so quickly and formed a lasting friendship. He was a country boy too. Macon, Mississippi, a town of a bit more than two thousand, was blessed with Gene's singing in the church choir. He had a voice and song writing talent that could have taken him far. In fact, he took a leave of absence from the National Weather Service and went on the road for one year with Mel Tillis, writing songs for him and warming up his audiences.

We sat in my office discussing things we had in common and bragging about our kids. I showed him a photo of Carrie. She was two, going on three, and had long blond hair. After seeing Carrie's picture, Gene abruptly excused himself and left the building. Within thirty minutes he returned with a guitar and piece of paper onto which he had penned a song that he entitled, "Daddy's Little Girl." He sang it for me. The song was very good. Looking back on that day, it is easy to understand why Mel Tillis took him on the road. But, after a year of accompanying Tillis and his band, Gene decided that he had to give it up for his family. I have no doubt that he could have "made it big."

After hearing "Daddy's Little Girl," I wrote my address on a Post-it sticker and handed it to Gene. "You have to sing this to Rheta," I said, almost begging.

I Called Rheta to let her know Gene was coming over. Within minutes, Rheta answered a knock on the door.

There stood a tall man wearing cowboy boots and holding a guitar. As the door opened, he exclaimed, "Hello Momma. I'm Gene Poag Jr and I'm here to sing a song for you."

Lay That Pistol Down Babe

Agnes passed too far west of the Florida Keys to have any adverse effect. After the storm departed the Keys, so did Gene Poag. It would be several years before we had occasion to see each other again. He moved around quite a bit as did I. We did so to gain promotions. I do not know where all Gene's career path took him, but we were both Radar Meteorologists at about the same time. As such, we had the grade GS Nine. Gene had long felt that Radar Meteorologists were under classified and therefore under paid for all that we had to know. In those days, the weather radar was the heartbeat of severe storm forecasts and warnings. The only weather satellite was a polar orbiting one. It would only give useful information for a given area twice a day. In between, there was nothing except radars. Somehow, Gene ended up testifying before Congress and did such a good job that the National Weather Service upgraded the Radar Meteorologist positions to GS Eleven.

My career path took me from Key West (a lateral move because Key West was a dead end) to Atlanta, Georgia where I worked as a Fire Weather Meteorologist and then to Oklahoma City with my next stop being Anchorage, Alaska. It was in Anchorage that I reconnected with Gene. Anchorage was the location of the offices of the Alaska Region and Forecast Office where I worked. Gene showed up at a conference one day in the regional headquarters. I ran into him at the elevators. I could not believe my eyes, but it was true. I was looking at the same tall lanky cowboy who had brought his guitar to my office years before and serenaded me.

I wasted no time inviting him to our house for dinner. He would be in town for four more days but had obligations for the next two evenings. We set the time to get together for a couple of days later. On the agreed upon evening we enjoyed a wonderful meal that Rheta had prepared and then insisted upon hearing Gene play and sing the songs he had written. No problem. No begging needed. Gene was in his element whenever he was plucking the strings of his guitar and singing songs he had written. I really think we would have had to hog-tie him to keep him from singing. He told us that on his flight from Point Barrow, he had entertained the passengers all the way to Anchorage and they loved the free concert.

Did I say Point Barrow? What was Gene doing there? Well, Gene always thought it was because he had gone over the heads of the Southern Region of the National Weather Service and testified before Congress. Perhaps it was. At any rate, for whatever reason, Gene had felt he would be at a dead end if he stayed in the Southern Region. So, having worked in Alaska some years before, he called the director of the Alaska Region and asked for a job. Stuart Bigler was the director then and he told Gene that the only opening he had was an "Official in Charge" of the little weather station at Point Barrow. "I'll take it," Gene told the director.

Gene was not at Barrow very long though. One day two men walked into his office. They were from the headquarters of the National Weather Service in Washington. One was the Deputy Director. His first words were, "So you're the guy that cost the National Weather Service all that money for grade structure upgrade?"

Without hesitation, Gene replied, "Hell, man, I'm already at the end of the earth. What else are you going to do to me?"

The Deputy Director answered, "We want you in Washington. Wrap up your affairs and as soon as we can get your replacement up here you will be on your way. But we will have to get you upgraded because the Director told us he does not have anyone under a grade of GS-fourteen working for him.

Later, after his replacement arrived, Gene stopped by to spend his last evening in Alaska with us. He would be on his way to D.C. the next morning. That evening was a wonderful memory maker. Rheta popped some popcorn and we sat in the living room and listened to one song after another that Gene had written. His song writing talent was amazing. He only published one album which he entitled "The Fightin' Little Judge," and included the title song about George Wallace of Alabama who had had his run for President interrupted by wounds incurred in an attempt on his life.

I was working the graveyard shift that night and had to be on duty at midnight. At around eleven thirty we said our goodbyes and Gene left for his hotel when I left for work. About fifteen minutes after we left, Rheta heard a knock on the door. She got Steede, our teenaged son, out of bed. She opened the bedside stand and took out the little Mauser three-eighty pistol and racked the breech back to put one in the chamber. It jammed.

Nervously, she called out, "Who's there?"

The answer came back. "It's me, Gene. I forgot my briefcase."

She opened the door. Gene walked in and saw her standing there with the bullet protruding sideways from the ejection port of the gun. I was not there to see the sight, but I bet it was a toss-up as to who was most frightened.

"Momma, put the gun down and I'll take care of it."

Neither of us ever saw Gene again. I wondered what happened to him. One day I googled his name, knowing that the internet would have a record of him if for no other reason than that he had testified before Congress. I did find him but not the way I had hoped. What I found was an article in "Weather Wise" magazine containing his obituary. The storyteller, songwriter, and dear friend would sing no more. Well, at least he would not sing down here. Gene was a believer in Jesus, so I know he is singing God's praises in heaven.

I never think of our last evening together, but what I'm reminded of is that song "Pistol Packing Momma," and the one refrain, "Lay that pistol down, babe. Lay that pistol down. Pistol packing Momma lay that pistol down."

Never Bear Hunt with a Three-Eighty

Two events stand out in my mind when I remember Denali. Camping at the foot of the great, awe-inspiring mountain is one such event. A tour bus ride high up on the mountainside is the other. Denali is the Koyukon (Native American) name given to Mount McKinley whose tallest peak is twenty thousand three hundred and ten feet. The Denali National Park and Preserve occupies over forty-six hundred square miles of interior Alaskan wilderness. The mountain itself is indeed a sight to behold but it hides many things from those who merely view it from a distance. Either by researching the subject of Denali or by information given by a tour guide one learns of some of those obscure events, such as attempts to climb Denali. The first person to try it in nineteen-aught-three failed. His name was James Wickersham. The first person to claim he had conquered the peak was Frederic Cook but it was later proven that he did not. In nineteen-thirteen, several men ascended the south and west slopes and succeeded.

Camping at the foot of the great mountain, Denali, was one of those experiences you are glad you had but never want to have again. Well, let me qualify that. Rheta would never want to have it again. We found a beautiful spot to set up our tent. It was on a wide bank of the Little Susitna (Little Su) River. To the west, the river bank narrowed and merged into a small grove of pussy willows. Of course, later we learned that willows are a favored haunt of bears.

"Okay, kids, get the tent poles and stakes out of the station wagon and I'll get the tent," I instructed.

Steede was thirteen. Carrie was eight.

It was Rheta's first time to stay in the tent. I lost count of how many times I had spent the night in it. You see, it was the same tent that I have written about several times. It was Henry's tent and would comfortably sleep six. Many a night this old canvas shelter served my dad and I, along with Henry, Uncle Press Squires, and occasionally other of dad's friends. Henry had given me the tent at the end of my senior year in high school. I took it to Antonito, Colorado where I worked for my Uncle Johnny bucking slabs at his sawmill.

Oh, if that tent could have talked. For sure, it would have told of the prank my cousin Benny Dobbs played on me. He pointed out some bear tracks near my tent earlier that morning (I found out later that he had made them with his hand) and then culminated his hoax by leaping on top of me as I slept in my tent that Saturday afternoon. I rewarded him by letting out a blood curdling scream. I should not have done that. He did not deserve to be able to laugh and tell that story which I am sure he would do if he were here today. I swore to get even but I never did.

But, now it was Rheta's turn to be initiated. It was her first campout. She hated everything about it. I know she did not sleep a wink that night. We had eaten supper on the way up to Denali and it was getting late. We got the tent set up and I sent Rheta and the kids to get some wood for our fire the next morning. When they returned, I got an ear full about how scary it was what with bears in the area.

"But, Rheta, it's broad-open daylight, and bears won't bother you if you make lots of noise. In Alaska, folks wear "bear bells" to let the bear know they are around. You never want to surprise one of those rascals."

While it is true that it was daylight, it was, nevertheless very late by the time we got set up for the night. As we prepared for bed, I looked around and lo and behold, Rheta was gone. Oh no!

Surely, she has not gone down river by herself. Then I saw her. She was crawling into the Ford station wagon.

"Rheta, what are you doing?"

She turned around and stared a hole right through me.

"What? What did I do?"

I wondered if this was a new way of sleeping on the couch.

"If you think I am sleeping out here with the bears, you are crazier than a loon," she said, as she pulled her three-eighty (sometimes called a nine-millimeter short) from her bed roll. "Dear, shooting a bear with that thing will have no more effect than if you shoot it with a BB gun. Besides, we can't take that thing inside the park."

Nevertheless, after a brief attempt to get Rheta to change her mind and sleep in the tent, I gave up. One thing I had learned from living with a Type A personality was, "Don't mess with the Rheta," when she has made up her mind. As she climbed into the station wagon and pulled the covers up around her neck, the kids piled in too. Guess where that left me. Give the reader a prize. My mother didn't raise a fool. I swallowed my pride and reluctantly changed my sleeping plans. I got into the wagon, wrapped myself in a blanket, partly to hide my shame, and went to sleep. After all, you should never bear hunt with a three-eighty.

Stop the Bus, We're Getting Off

Having largely spent a sleepless night, what with all of us cramped up in the Ford station wagon, we rubbed our eyes and crawled out of our sleeping bags to face an uncertain day. Had a congregation of bears gathered during the night and patiently waited for morning to interrupt our breakfast and have theirs? If bears held off on molesting us, we would have our breakfast of sweet rolls and donuts we had picked up before we started. Rheta and I would have coffee and the children would have bottled juice. The bears had taken pity on us. I suppose they did so because it was summer, and grass was tall and green, and they love grass. Their bellies were probably full and, besides, they had probably been told, "don't mess with the Rheta." For whatever reason we were spared, we were grateful. We dismantled the tent, packed it up, moved to a campground, and set up the tent again. Shortly, we would take a tour of the great mountain.

We arrived at the bus tour departure point which was the Visitor Access Center. All of us were excited about the ninety-mile all day tour. According to the brochure, we would see our fill of park animals. There would be an abundance of caribou, moose, wolves, Dall sheep, and grizzly bears. Did I say grizzly bears? We were reminded again of how close we came (or thought we came) to being bear breakfast or a midnight snack. We were going to travel deep into the Kantishna Wilderness on a road that barely had enough room for the tour buses to pass. We did not know about that. After our ordeal on the Alaska-

Canadian Highway, none of us wanted any more dangerous mountain slopes. But here we were on a road where it was necessary for both buses to challenge the margins of the road. By challenge, I mean that it was necessary for the driver to literally put the wheels of the bus on angel's wings to avoid dropping a wheel off the edge and then rolling down several thousand feet.

However, the moment we got on, we should have expected that this might be a ride from hell. It would be hard to find a ruder bunch of individuals. The first rattle out of the box a woman on the opposite side of the bus, the upslope side, got up from her seat, came over and, leaning over us, opened our window, informing us that she could now see. That act set a precedent. From that point on, every time someone on our side spotted an animal, everyone on the other side would rush over, lean against us, and engage in various utterances. After, our bus appeared to be on the verge of running off the road a few times, we were ready for the trip to be over. The rampant impoliteness of tourists lent urgency to our desire. We began to see people standing on the side of the road being picked up by returning buses. Steede said that he wanted to get off.

Rheta responded, "No, we are all getting off. I have the keys to the car. Anyone who wants to go back to Anchorage had better speak up now."

I spoke up. "Driver, stop this bus. We are getting off."

Don't Try to Make Little Bear by Bedtime

While we were living in Key West, we drove to Texas and Oklahoma a couple of times to see our family. Driving straight through took about twenty-three hours and took us through the fringes of the Okefenokee Swamp in southern Georgia. We decided to drive straight through. We could save on lodging and be able to spend more time with our family.

Long stretches of the route took us through swamp land, ripe with large cypress trees that were covered with Spanish moss. Now that was a very eerie sight. It was spooky, I tell you, spooky. At four in the morning after long hours of driving, our eyes began to play tricks on us. "Was that a zombie standing by that big cypress back there? Look! There is another. We need to stop for a rest somewhere. I am starting to see things."

"I saw a sign a mile back that said, 'Little Bear,' maybe we can get some coffee and pop for the kids." In that part of the country, soft drinks are called pop.

Half asleep, I drove on. I suggested to Rheta, "I must have misread the sign back there. I'm pretty sure we have driven more than seven miles."

She agreed. We drove and drove.

After some time, Rheta said, "Look. Little Bear five miles."

"Wonderful, I sure need a restroom break."

We drove much farther than five miles. Still no sign of Little Bear, not even a glow of distant lights. "You know, I think Little Bear must be so small a place that we missed it. I do not think we are too far from Valdosta. Why don't we give up on the idea of driving straight through? Let's get a motel in Valdosta." Rheta agreed. The kids did not mind. Their motel was the back seat. Now perked up a bit by the anticipation of a soft bed, we were more awake and eager to get to Valdosta. About that time, we saw the sign again. We both saw it clearly. We both were silent. Neither of us wanted to admit that we could not read. We could not see in the dark, but I am sure our faces were as though we were badly sunburned.

The sign said, "Litter Barrel one half mile."

If You See Eyes Under the Hen, Come Back Later, Much Later

I love chickens. I always have. Momma always had lots of chickens when we were growing up. In the spring, she would order baby chicks from Sears and Roebuck. Mail order was big in those days. You could order pretty much anything from kitchen appliances to farm equipment and everything in between, including all kinds of fowl.

There were options when ordering hatchlings. It was possible to order straight pullets (females), but Momma always took her chance and ordered unsexed lots. They were cheaper. We ate the roosters when they got big enough. She always tried to keep one rooster to fertilize the eggs. That way the chicken crop would be replenished, and we could harvest the roosters (fry them when young and bake them when they got old and tough). Old hens do not lay many eggs, so we also ate them.

Our barn was typical of barns in the nineteen-forties. The building had a staircase leading up to the loft. We stored bales of hay there. A hallway ran the length of the building and had cribs, stalls, and storage rooms on each side. One such room was a long rectangular room with a wooden floor. We stored the yearly corn harvest there. Dad's corn crop would fill that large crib full. We fed whole ears of corn to our pigs and "shelled" the corn for chickens. The hens nested in boxes attached to one side of the barn.

My youngest sister and I were responsible for feeding the

chickens and gathering the eggs. Sometimes when we would gather eggs, a hen would not want to get off the nest and she would squawk and fluff her feathers up to scare us away. By her behavior, we knew that she wanted to be a mother. So, we would take a pencil and mark up a dozen eggs or so and place them in the nest. She never objected to more eggs. She just did not want us to take any out. Marking the eggs made it was easy to tell which were fresh eggs. The hens competed for nesting boxes, so when the setting hen left her nest to get food or water, some squatter would steal the nest and keep it until she made her daily deposit. Twenty-one days later, we would have a few more chickens, albeit, baby chicks.

Sometimes when we gathered eggs, a hen would be occupying the nest so that we had to reach under her to collect the eggs. If she had not yet laid her egg that day, she would be hesitant to get off the nest, causing us to reach under her to collect eggs that other hens had laid in that nest. When we had to "rob" her nest when she was on it, we were very careful to determine what was under her. There might be a chicken snake having a feast beneath her.

Each time we spotted a snake under a hen, we told Dad and he would reach under her and get the snake at the base of its head and pull it out from beneath the hen. Then he would take the snake by the tail with his other hand, walk out into the barnyard, drop the head, and pop the snake like a whip. Off would come the head and however many eggs the reptile had eaten, usually one or two, would splat against the side of the barn.

When we moved back to Springtown and built a house on our acreage, I built a chicken house and yard. I promptly purchased some baby chicks, all pullets. I worked shifts then and did not want a rooster interrupting my sleep. Carrie was in middle school then. She loved the chickens. It was inevitable that she would eventually encounter a chicken snake. Few things startle me more than a snake which I accidentally stumble upon, but once I get my eyes on it and back away to a safe distance, I am okay to assess any potential danger.

I have been known to capture the non-poisonous ones and keep them in an aquarium for a day or two for my son to play with. Well, that was before Rheta caught me. Any time she spotted one she would call out, "Don, come kill this snake." She accuses me of always responding with, "is it poisonous?" "Don, I don't care whether it's poisonous or not, just kill it please." She is deathly afraid of bugs. She once jumped from our truck as we drove slowly across a field when a grasshopper flew through an open window and landed on her. Snakes, up close and personal, will give her tachycardia. One day I brought a little garden snake into the house and showed it to her. I thought she was going to divorce me.

I knew it was bound to happen and it did. One day, Carrie went out to play with the chickens but saw something that frightened her. She came running into the house exclaiming, "Dad, I saw eyes under a hen." I knew immediately what it was even though she did not.

I told her to follow me. We went into the chickenyard and I promptly reached in under the hen and latched onto the snake, dragging it quickly from under her. There was only one problem. I grabbed the old serpent four of five inches below its head. In short accurate bursts like the staccato of a machinegun, it began swinging its head from side to side, each time sampling each knuckle on the back of my right hand.

Did I tell you that I dropped that rascal like a hot horseshoe? Well, I did. Then, with blood oozing from each knuckle, I turned to Carrie and said, "Daughter, that's not the way my dad did it so the next time you see eyes under the hen, come back later, much later.

Civilian Once More

The place was Carswell Air Force Base in Fort Worth, Texas. At that time, Carswell was a Strategic Air Command (SAC) base. I was just twenty-two years old as I stood at the desk of the commander to get my re-enlistment talk. I will not mention his name, but he was a Lieutenant colonel commanding the twenty-second detachment of the twenty-sixth weather squadron. He was a hard-nosed World War II glider pilot who had replaced Colonel Higgins about a year before. I am here to tell you, that was a real comeuppance. My previous commander, also a lieutenant colonel, was a jolly man who loved to talk sports with his troops. On payday, we would walk in and ask him for our check. Often, he would ask us who would win the game, the Giants or the Dodgers. He was a Dodgers fan.

The new commander, though, was a real stickler for protocol. I walked in, saluted, gave my name, rank, and serial number and then stated that I was reporting for pay. So far so good. Then came the fatal mistake. I turned around and walked out with my check in hand. Before I got to the door, I heard a yell. "Mankin! Get back here and do it right this time." "Sir! Yes Sir," as I returned to his desk, squaring all corners. This time, as I made my hasty exit, I did it right.

His desk was offset so that it was necessary to walk away from his desk, do a right flank, take a few steps, followed by a left flank, and then march forward. I wonder if he positioned his desk like that on purpose. That rascal commander! I bet he did. As I walked to the door, I heard him say, "Airman Mankin, come back here." Oh no! What did I do wrong this time? "Mankin, you would never make a good soldier. You might make a good Airman someday." What an insult! Biting my tongue until it almost bled and choking on my words, I replied, "Sir, yes sir."

A month or so before I was separated from the Air Force, the commander called me into his office. "Airman Mankin? "Sir. Yes Sir." He said, "I'm looking at your file and I see that you don't have a college education. You will never amount to anything on the outside" (as though I was in prison). I must confess, sometimes I thought that I was. At any rate, he was wrong about me and I would show him. I knew something he did not. I had already interviewed with the director of the Southern Region of the Weather Bureau (now, National Weather Service). Furthermore, I was hired on as a GS-5 Meteorological Technician (glorified name for weather observer). My salary would be four-thousand-and-forty dollars a year. To a kid whose paycheck was one hundred and twenty-four dollars a month, such remuneration was all the money in the world.

As it turned out, after six and a half years of study, mostly night classes, I completed my degree in mathematics and physics. Once I had obtained classification and title of Meteorologist, I joined the local chapter of the American Meteorological Society which met at, of all places, Carswell Air Force Base. My old commander, also a member of AMS, was there as was a Lieutenant I served under. The difference this time was that I never squared corners. I never saluted. I never stated name, rank, and serial number. In fact, I never asked the commander for my check. I walked in, head held high, greeted him, and made small talk. I could do that now. I was a civilian once more.

ALERT, Prankster on the Loose at Amon Carter Field

His name was Aurel J Knarr and he was Meteorologist in Charge of the Weather Bureau Forecast Office at Amon Carter Field. He was a godly man, a true pattern for any young man to follow. He was my first boss in the weather business and I admired him more than anyone I ever worked for. I never once heard him so much as say "damn." He demanded quality work and understood his role to be that of keeper of taxpayer dollars. He was as thrifty with public funds as he was with his own money.

We were required to record weather observations on the regulation WBAN10a and 10b using a special pencil. On everything else we could use an ordinary 2H pencil. He would leave the 2H pencils in the supply cabinet that was unlocked, but, the special pencils were kept in a safe next to his desk. When we wanted a new pencil, we had to let him inspect the one we were using. If it was of sufficient length to get a bit more use, we were told to come back when it was more "used up." That practice evoked impatience among some of my fellow observers, but I admired him for it. The thought of disappointing him motivated me to excel. But, at heart, I was born a prankster.

I began on-the-job training at being a mischief-maker early in life. As a kid, I delighted in playing tricks on my dad. I had a little battery-operated transistor radio which was one of my trouble-making tools. The interior of our house had single clapboard walls. Radio signals had no trouble penetrating them. In the evenings, dad would turn on the big floor-standing radio in the living room and listen to his favorite program. I would lie on my bed on the other side of the wall and slowly change the station dial until I achieved signal interference. Once I heard the resulting whistle, I adjusted the station selection knob ever so slightly, first clockwise, then counter-clockwise. The wavering, back and forth, frequency change would cause my dad to try to correct the problem on the other side of the wall by doing the same. I never did it long enough to try his patience though.

I carried my tomfoolery skills into adulthood. For the most part, I kept them from Mr. Knarr. Many of my best acts were performed in the office of the Weather Bureau or on the roof. Our instrument shelter was located on top of the airport's main building. Along the perimeter of the roof was a three-foot high wall. To the west, it was a long way down to the ground from the roof. On the east, however, the drop was only a few feet to the roof of a porch-like structure.

I was single then. One night, after church, I brought some friends to the office to show them where I worked and what I did. Naturally, such an opportunity would never escape a master trickster. I promptly took the group over to the west wall, cautioning them to be careful. As they peered over the wall and were watching the cars below, I suddenly yelled like a maniac and ran to the east wall. As I jumped to the roof below, one girl screamed and then with legs buckling, fell to the floor. I'm sorry I did that. No, I'm not.

On another occasion, same place, on a very cold December day with the temperature hovering around eight degrees and very windy, the weather observer on duty went out on the roof to read the wet bulb and dry bulb temperatures from which he would calculate the dew point and relative humidity. A couple of weeks before, I had broken my ankle and was wearing a walking cast. I waited until my co-worker was standing at the instrument shelter and then locked the door on him. The phone rang, and I got busy and forgot about John on the roof. Moments later I heard a horrible sound. It was that of a metal panel in the door being kicked out. Something told me I needed to get out of the prank-playing business. With my concrete leg, I ran like Chester on Gunsmoke, and opened the door to let him in. I didn't know I had that many names. Among other things, to him, I was a prankster on the loose at Amon Carter Field.

ALERT! Prankster on the loose at Amon Carter Field ... Part II

In the nineteen-sixties, while pursuing my degree, I took some management courses. Back then, two extremes could be used to management styles. On one end of the spectrum was the Theory X manager. These managers were hardnosed and believed, very strongly, that employees could only be motivated by money and fear of "the boss." On the other end was the Theory Y manager who believed, likewise very strongly, that employees were generally desirous of doing a good job and would be motivated more by opportunities than by money.

Aural J Knarr was more to the Theory X end of the scale. He would allow employees to seek opportunities but required proof, by performance, that they could succeed after taking advantage of a chance to advance. Under Mr. Knarr, it was necessary to have a set of top-notch work ethics if one wanted to "get ahead." Hard work was not a new concept for me. From my high school summer days working on county roads with a shovel and wheel barrow, I had given a hundred and ten percent to every job. As a skinny kid shoveling gravel with men my dad's age, I still remember being fussed at. "Donald, come sit down and rest. You're making us look bad." Of course, they were joking about being made to look bad. I think they secretly were glad I worked so hard.

Regarding work ethic, my guiding principle has always been "give every job your total devotion and show up early and leave late." For that reason, I think, Mr. Knarr always liked me and so when I asked him if I could attend college, he granted permission. In the weather business, it is usually feast or famine insofar as workloads are concerned. I was a Radar Meteorologist at Amon Carter Field and during slow periods I busied myself with my studies at my radar console. Work ethic aside, I always made sure that an opportunity for mischief did not pass me by.

It was the graveyard shift shortly after July Fourth. I was sitting in the weather observatory with the observer whose name will go unmentioned. I was working on my lessons. He had drifted off to sleep and it was nearing observation time. Meeting deadlines was an absolute must with Mr. Knarr.

I needed to awaken the observer, so he would get his observation out on time. What better way to roust someone than with a left-over fire-cracker. Yep, I sure did. I quietly reached into my pocket and took out one of two little firecrackers. I smoked back then so I pulled out my lighter and touched it to the end of that little powder packed rascal. I cringed as it began the characteristic hissing sound. I wanted the bang to wake him, not the hiss. He did not spoil my fun. He just kept on snoring. "Kaboom!!!" Dang! I should not have done that. The observer, jumped. The wheels rolled. The chair separated from its occupant. That was nineteen-sixty-two. I wonder if his ears are still ringing as mine are.

Where There is Smoke, Look for John

John was a weather forecaster at Amon Carter Field when I was an observer there. I reckon I have never seen a more intelligent man. He was brilliant and could have had a doctorate in physics but for the fact he just did not want to do a dissertation. John had more semester hours in physics alone than most degreed individuals have in total hours. "Wow! I've hit the mother-lode here.," I thought as I immediately determined to use John as a tutor and resource. Any time I need help with my college work, I will just ask John, I schemed. Yes, he was brilliant but oh how eccentric!

I carpooled with John. We both lived in the Poly area of east Fort Worth. One week I would drive my car and the next week he would drive his. On the days that I drove my car, I would cringe as I let John out at his house. He would get out of the car and slam the door with every ounce of energy he had. His car door was hard to close. Mine was not. I feared each time that my window would shatter as if hit by an eighteen-wheeler. Wouldn't you think that a guy with over a hundred semester hours of Physics might suspect that only so much force can be applied without breaking something? But, being the respectful young whippersnapper that I was, I dared not fuss at John. I would simply find a work around.

I knew what I had to do. The next day as he got out and drew the door wide to slam it, I leaned over and "stiff-armed" the door. Dang! I should not have done that. But, you see, I had not studied physics yet so consequently, I walked around for a week or so with an arm sticking out my back so that if I turned sideways I looked like a six-foot letter T. I never tried that again, but I did have the distinction of being the first Mr. T.

John wore a hearing aid. Hearing aid technology in nineteen-sixty left a lot to be desired. A shirt pocket device, about the size of a pack of cigarettes had a small rubber tube that ran up to a bulky and somewhat unsightly large half-moon-shaped ear piece. The volume control was located on the shirt pocket unit. John was a good-natured man in his late forties. He was forgiving of my pranks although I must have tried his patience at times, like when I messed with him about his hearing aid.

One day, I positioned myself in front of him, book in hand as though I wanted help with a problem. I began whispering so that he would see my lips moving but not hear my words. John's hand went immediately to his shirt pocket to increase the volume. I whispered again. Again, John reached for the volume control. Once more I repeated, only this time I spoke softly so that he could barely hear. Once again, this brilliant physicist adjusted the volume at which point I spoke as though I was raising my voice to accommodate him. Dang! I should not have done that. I could tell it must have sounded like I was using a bull horn. I was never able to pull that prank again.

Perhaps one of my better shenanigans was the fax paper incident. I have mentioned before that the central office in Washington D.C. would send weather maps to field offices via facsimile. In the field office, a bell would ring to alert personnel that a fax chart was coming. We would rush over and push a button to synchronize the incoming chart. If we did not, the map would be out of phase and it would be necessary to cut and paste it to see weather patterns as they should appear. The fax paper was treated with a chemical that would imprint images when a burning stylus moved across the paper. The smoke would waft its way through the room, making for an unpleasant odor for a few minutes. The perfect prank came to my mind.

John chained smoked and always flicked the ashes from his cigarette into a trash basket which usually contained old charts wadded up. The forecasters prepared their forecasts while sitting at a long desk with a high bulletin board-like back onto which weather maps were posted. On more than one occasion, John had set the trash can on fire. Once, he flicked his cigarette into the can and caught another type of paper on fire. He jumped from his chair, headed for the window, opened it and was hanging out the window trying to extinguish the blaze when Mr. Knarr walked in. Boy howdy did he get reamed out.

So, a few nights later, I slipped around behind the desk, lit one of the fax charts and let it smolder. In about a minute, I heard John exclaim, "Oh my!" I snickered to myself as he jumped from his chair and headed for the window. Just as he was about to hang out the window again, he realized that there was no fire in the trash can. It was just that prankster kid at work again. We all got a laugh out of it, even John. This time the general rule failed. But, we kept the rule in effect because we knew John would again set a basket on fire and we could invoke it once again, where there is smoke, look for John.

I Can't Think About Jesus

I can't think about Jesus without remembering that He is God. "In the beginning was the Word (Jesus), and the Word was with God, and the Word was God." John 1:1. "And the Word became flesh, and dwelt among us, ..." John1:14.

I can't think about Jesus without remember that He created all things. "All things came into being through Him, and apart from Him, nothing came into being that has come into being." John 1:3.

I can't think about Jesus without remembering why He came to Earth and dwelt among us. He came "...to seek and to save that which was lost." Luke 19:10. God created man with a free will. He first created Adam and Eve and set them in the middle of the beautiful, unspeakably majestic Garden of Eden. "The Lord God planted a garden, toward the east in Eden; ..." Genesis 2:8. God assigned the responsibility of maintaining the garden to Adam and told him that he could eat anything in the garden except the fruit from the tree of knowledge of good and evil. He warned Adam that if he ate from that tree, he would surely die. Genesis 2:17.

I can't think of Jesus without remember that, according to the bible, it was the sin of Adam and Eve that, as part of the curse God placed on them, is passed down through generations to us. Romans 5:12. By that one act of sin on the part of Adam, all are born with a condition of sin that will, without fail, bring fruits of actual personal sins which will condemn. "The wages of sin is death, but the gift of God is eternal life through Jesus Christ our Lord." Romans 6:23.

I can't think about Jesus without seeing the futility of my existence apart from Him. I must have His Grace. I must have it because it is "by grace you are saved through faith and that not of yourself. It is a gift of God, not of works lest any man should boast." Ephesians 2:8-9. We can't work hard enough to cause God to accept us. The only way to do that is through faith in His son, Jesus. Jesus said, "I am the way, the truth, and the life. No man comes to the Father except through me." John 14:6.

I can't think about Jesus without remembering His agony of the Cross. God, the Father had sent God, the Son into the world to save us from our sins and make us acceptable to Himself. Jesus lived His life on Earth absolutely without sin. Because of His perfection, Jesus was the only possible sacrifice to atone for our sins. So, the Father," ...made Him who knew no sin to become sin on our behalf, so that we might become the righteousness of God through Him." 2 Corinthians 5:21. So then, when Jesus died on the cross, the Father put all our sins, past, present, and future, on Jesus and punished Him in our place. Then the Father transferred the righteousness of Jesus over to our ledger, thereby making us acceptable to Himself.

How do we set this wonderful accounting procedure in effect so that God's bookkeeping will result in eternal life for us? I can't think about Jesus without remembering that all one must do is to accept God's gift of eternal life. We do this by casting ourselves with all our sins upon His mercy. We must confess and believe in Jesus. "That if you confess with your mouth that Jesus is Lord, and believe in your heart that God raised Him from the dead, you shall be saved." Romans 10:9

I can't think about Jesus without thinking about how wonderful it is to have eternal life which began for me the moment I put my trust in Jesus. I can lay my head on my pillow at night knowing that should I die before I wake, I will be with Him.

Praise the Father, Son, and Holy Spirit that I can think about Jesus.

Momma, I'm Cold

"Momma, can we stop and rest?"
"No Son. We have a long way to go and must hurry to get there."
"Momma, where are we going? I'm tired. I want to rest."
"We must keep going, Son. No time for resting now."
"But, Momma, why are we in such a hurry?"
"Son, we must get to the forest before the big light in the sky disappears."
"Momma, what will happen then?"
"It will get dark and cold. We must find shelter."
"But, Momma, I'm very tired."
"Hush, Son. You are starting to get on my nerves. When you grow up you will understand."
"Okay, Momma."
"Momma, when I grow up, will I be big like you?"
"You will most likely be bigger than me. You are a male."
"Momma, can we stop and rest now?"
"No, Son. We are almost there. Then we will have a long time to rest."
"Momma, where have we been?"
"Son, we have been in a place called the Alaskan wetlands. We had lots to eat there, but now we must prepare for winter. We will change drastically soon."
"Momma, I'm scared."
"Don't be afraid, Son. It is only frightful the first time, but the one who made us also made us so that we will survive the

winter."

"Momma, who made us?"

"Well, son, humans call Him God and He cares about every creature He made, be they great or small."

"Momma, did you say we are wood frogs?"

"Yes, son. Burrow harder. Wood bark, dirt, and leaves must well cover us. This is where we will spend our long winter night."

"Momma, if we can't stay in the wetlands, what will we eat?"

"My dear little boy, you have so much to learn. God made us so that we will not need to eat while we are here."

"Momma, what will it feel like? Please tell me about it."

"Son, move close to me and I will tell you. It is a very complicated process. Humans say that cryoprotectant chemicals keep us alive when we are frozen, but it is important to remember that no matter how it is done, God did it."

"Momma, does it hurt?"

"No, it does not."

"Momma, tell me what will happen to me please."

"Okay, but then we must sleep. It is almost time. Move closer to me. The night will get colder and colder because the big light in the sky has gone out for a long time. The first sign that the time has come will be that your eyes and legs will freeze. Then your entire body will freeze solid. You will not move. You will not even breathe. In fact, even your heart will stop beating."

"Momma, are you happy that we live in Alaska?"

"Yes, I am very happy."

"Momma, do you ever want to be big like the brown bear and the moose?"

"No, I do not because God made us this way and He watches over us and wants us to be content with whatever is our circumstance."

"Momma, I love God."

"I love Him too, Son. Now be quiet and sleep. You will fall asleep and not awaken until spring when the sky will once again be visited by the big ball of fire."

"Momma, tell me about the light in the sky."

"Not now, my boy. I am sleepy."

"Momma, my eyes and legs feel funny. Momma, I'm cold."

Of Thongs and Things

A few years ago, our children were involved in their own activities on Thanksgiving Day. Since we had celebrated the holiday early, Rheta and I took advantage of the opportunity and went to Montreal unsupervised by our kids. The November air was crisp as one might expect Canada to be in the fall. The snow was falling ever so lightly as we left the snug little café where we had enjoyed a wonderful breakfast. Rheta wanted to shop in underground Montreal. So, we did so.

Underground, the shops were amazing as were the eatery choices. For me, it was a wonderful experience. There are few things that Rheta enjoys more than shopping. More than once she has outlasted our daughter-in-law and daughter when they go to the mall together. She literally believes in the expression "shop 'til you drop." Sometimes she will spend a couple of hours in the stores and go home with nothing.

"Rheta, how can you spend all that time and not buy anything?" I inquired.

The reply comes back, "Well I didn't see anything I liked."

I shook my head and said, "Okay then, let's go back to the hotel."

I bet you are thinking, "Well they went directly to the hotel and rested." Wrong. We window shopped as we strolled hand in gloved hand on our way back. Since only an occasional snowflake drifted down from a low cloud deck, we were in no great rush to reach our hotel. We continued to window shop. Momentarily, something in a store front caught our eye. Two mannequins were attired in the strangest looking things that rendered them almost naked.

"What in the world are those things," I asked.

"Those are thongs," Rheta replied.

"Things? I know they are things but what kind of things?"

Rheta explained, "No, I said they are thongs."

By now I was very curious. "Montreal sure has some strange things."

"Don, I told you. They are not things, they are thongs."

"Oh, I know it Rheta. I am just giving you a hard time."

As we stood and stared at the models behind the window, we began giggling like two little kids. The same thought crossed both of our minds at the same time. "Let's buy a matching pair for our friends." (names withheld). The more we thought of the look that would be on their faces when they opened the package, the more we giggled uncontrollably. But alas, our better judgement took over and we abandoned our plan and walked on toward our hotel. Still, one of us would chuckle to ourselves and the other would break into a full-blown hoot as once again we thought of thongs and things.

If You Eat At La Madeleine's, Learn French

A few years ago, I had been to see my eye doctor in Irving. She dilated my eyes. I hate it when they do that. It always makes me feel like I'm going blind. You would think that the doctor would realize their job is to prevent blindness, not cause it. But no, I think they get some sort of weird pleasure out of watching me stumbling around in the waiting room. But I always get my revenge. You would be surprised at how many people get up and leave before they see the doctor when I start falling over furniture and feeling for a chair while announcing, "I'm blind. I'm blind."

On this day, it was noon when I got out of her office. I was hungry, and La Madeleine's restaurant was just across the street. I managed to get across the street and walk thirty yards or so to the restaurant. It is surprising how much money panhandlers can make if they pretend to be blind.

I never fail to get nervous anytime I visit a doctor or dentist. I suppose I am the most nervous when I visit my urologist. At any rate, when I get nervous I feel a need to find the men's room. As hungry as I was, I needed to take care of more urgent business first. We eat at La Madeleine's in Grapevine on occasion and so from rote I found my way to the restroom. Apparently, no other men in the restaurant had the same necessity that I did because I was the only one in there which I thought a bit unusual since the place was full of lunch-goers. Even more strange was the fact that there were no urinals on the wall. I scratched my head in amazement and went on into one of the stalls.

I heard movement in the stall next to me but didn't give it much thought. Well, at least I didn't until the manager stuck his head in the doors and said, "Sir, you are in the women's room." Someone outside had seen me go in and advised management. "Sir," he said in a not very kind tone, "I know the words on the door are in French, but you should be able to tell by the figures on the door that this is the women's room." You know, I am pretty sure that nobody has ever left that restaurant as fast as I did. No, I have not been back since. Furthermore, if I ever eat at La Madeleine's again, it will be after I learn French.

Don't Stare Too Long at a Barracuda

In the early nineteen-seventies, Key West was a laidback little town at the west end of the Florida Keys. By laidback, I don't mean it was not a place of interest. Quite the contrary. The little four-mile by two-mile island has been host to many notable visitors. One author, Ernest Hemingway, became a favorite adopted son. In fact, he purchased and lived in the Hemingway house on Whitehead Street for eight years. I believe the onetime home and, later museum, is now closed. But, as I understand it, someone still cares for the many six and seven toed cats that have resulted from many years of in-breeding going back to Hemingway's pets.

Key West was also a favorite haunt of President Harry Truman. It is where he decided to designate a house at 111 Front Street as "The Little White House." The house had served as a naval command post through three wars as far back as the Spanish American war. Truman went there often, and his staff would fly down from Washington to meet with the President and conduct national affairs.

Another frequent visitor to Key West was Jimmy Buffet. When we were there in the early seventies, Buffet was unknown to anyone who did not frequent the local bars and nightclubs where he played. No one had yet heard of Margaritaville. Buffet had tried to make it big in Nashville as a country and western singer but had not yet achieved fame.

Since I didn't frequent the bars and nightclubs, my interest was focused on fishing. In my opinion, there was no better place for it than the Keys. I fished with a friend whom I had known from our Air Force days. He had a small Jon Boat with a motor. We confined our fishing to inside the reefs where average water depth was only about seven feet. I also loved to spear fish.

It was during a spear fishing outing that I met up with my first barracuda. Until then, the largest fish I had ever seen was a thirty-two-pound blue channel catfish that my dad and I took off a trotline on one of our fishing trips with Henry. But I was getting ready to change that. I was about to come face to face with a barracuda as long as a Rock Island freight train.

On a morning that we both had off, we gathered up our spear guns, snorkels, and fins and climbed into his Jon boat which was anchored in the canal at the back of his house and headed out to sea. The gulf stream, an ocean current, runs northward along the east coast of south American into the Gulf of Mexico where it turns east past the southern tip of Florida. The current then moves north along the east coast. It can be rather strong. It is called an opposing current because the prevailing easterlies move from east to west and clash with the gulf stream. This wind and ocean interface can set up some very uncomfortable seas. I had experienced one of my worst cases of sea sickness a month or so earlier. On that trip, I had thrown up in the water and the waves carried my stomach contents right back into my face. I wanted no more of that.

Since I routinely prepared a marine forecast for the area, I knew that such conditions would not plague us this time out. We jumped into the water and began looking for big fish to spear. I had been fishing for about ten or fifteen minutes when I suddenly spotted a barracuda in front of me. I had always heard that they are not aggressive and will not bother you unless you are wearing a shiny object such as a necklace or watch. I was wearing neither, so I was not worried. I kept fishing. The big fish moved along side of me for a short distance and then maneuvered into a position facing me. The more I looked at that thing, the bigger it got. Soon I began to think I was looking at a submarine. I had heard that barracudas will sort of shake their bodies before they attack. This one shook. That was getting too scary. I decided I would start back. I raised my face up out of the water to look for the boat. It was a small spot in the ocean.

With face in the water and looking behind me, I began to make my way back to the boat. My new friend was following close behind. After a while, I turned my face forward to swim faster. As I turned, I found myself eyeball to eyeball with another even bigger one. I poured the coals to it. If that vicious predator was going to eat me, then he was going to have to eat a coward. Now where would his bragging rights be to that? I guess he and I were thinking the same thing because he and the other one joined up and went off in a different direction.

I climbed into the boat, having learned a good lesson. Don't stare too long at a barracuda.

Hidden Sins

What does one think of when they think about hidden sins? I would venture to say that most people will think about sins they have committed that they think nobody has discovered. Certainly, that line of thought is not unreasonable, at least on the surface. We live in a world where the line between right and wrong, good and evil, morality and immorality, and what is true and not true has become so blurred that our sins are not too difficult to hide so long as they are just "little sins." Nobody cares.

Ah but someone does care. By thinking otherwise, we delude ourselves. God cares. God sees all. Nothing is hidden from Him. "Nothing in all creation is hidden from God's sight. Everything is laid bare before the eyes of him to whom we must give account." Hebrews 4:13. In fact, God cared enough about sin and its destructive effect on every human life that, before He laid the foundation of the world, that He had a plan to deal with it. "For God so loved the world that He gave His only begotten Son that whoever believes on Him should not perish but have everlasting life." John 3:16

Jesus Christ, the living son of God, can redeem the vilest soul. One can't sin so much that Jesus can't save the sinner. Jesus said, "...Him that cometh unto me I shall in no wise cast out." John 6:37. He didn't say that only those who were perfect would not be cast out for "there is none that is righteous, no not one." Romans 3:10. No, Jesus said, "Come to me all you who are burdened and weighted down with care, and I will give you rest." Matthew 11:28. What greater burden does one carry than sin itself?

The prophet Micah told of Jesus' coming to Earth. "He will turn again, he will have compassion on us; he will subdue our iniquities; and thou wilt cast all their sins into the depths of the sea." Micah 7:19 The bible is clear that all who are not believers in Jesus Christ are enemies of God because we are born with a sin condition that came from Adam. But for anyone who accepts the free offer of eternal life through Jesus Christ, God the Father said, "Can a mother forget the infant at her breast, walk away from the baby she bore? But even if mothers forget, I will never forget you, never." Isaiah 49:15.

Speaking of those who trust Christ, Psalm 103:12 says, "He will remove our sins as far as the east is from the west." So, while it may be true that one can hide their sin from people, they can't hide their transgressions against God from Him. Trust in Christ. Have your sins removed from you as far as the east is from the west. Let God hide them from Himself in the deepest part of the ocean. Then and only then will you truly have hidden sins.

Wilburton Main Street on Saturday Night

It was nineteen-fifty-three in Wilburton, Oklahoma. I had just gotten a driver's license. I failed my first test because I did not see a stop sign. Weeds had grown up around it and it was hard to spot. I blew right past it, dumb, fat, and happy. Oh, well, make that dumb and happy. I was a very skinny kid. The Oklahoma Highway Patrolman failed me. However, on my next attempt, the same officer pointed to the stop sign and reminded me that I failed to stop the last time. I guess he felt sorry for my having failed because a stop sign that was almost covered with weeds. This time I stopped. This time I passed.

At age sixteen, a whole new world had just opened for me. My mode of transportation would no longer be old Snip although I appreciated him for the many miles I put on him while delivering the Grit newspaper. I would still use him for that until I could have my own car.

Dad was very good about letting me drive his car. It was a brand spanking new fifty-four emerald green Chevy Belaire. I put the first scratch on it. I had driven it to Les Mitchell's house where I joined the other kids at the swimming hole on Little Fourche Maline creek that ran through Les' hay meadow. It was a stretch of water about twenty yards long and ten to fifteen-feet deep. An elm tree on the bank had a large limb overhanging the deepest part of the swimming hole. We often went there to while away the hours, climbing up in the tree and diving or jumping off into the cool clear water.

On this day, when I got into Dad's car to leave for home, one of the kids grabbed hold of the bumper and ran behind me. I leaned out the window to shout at him and ran the car into a ditch, thereby marking it with an abrasion about an inch wide and four inches long. Fortunately, it was very superficial, and I was able to buff it and almost completely remove it with Turtle scratch remover. Dad never noticed. At least if he did, he never mentioned it.

I would drive that car many times before getting my own nineteen-fifty-two Studebaker with Dad's help and my summer earnings from working in one of the two grocery stores in Wilburton. I did not start dating until my senior year but guys without dates would engage in a well-established ritual in Wilburton on Saturday nights. We would drive slowly through town, make a u turn and drive back through town. Gas was twelve-cents a gallon then, so we repeated that little back and forth until we either found a girl who was unattached or hooked up with a group at The Green Frog café or one of the other two eating places in town.

As I reflect upon life as it is for young people today and what it was like for us in fifties, I still must view those days as excellent. We did not have cell phones. We did not have computers. Heck, we didn't even have air conditioners in our cars, but we did have lots of fun. After all, we were cruising Wilburton Main Street on a Saturday night.

"Holy Horrors to Hell"

If I recall correctly, in the nineteen-sixties, the University of Texas at Arlington, my alma mater, had very few classrooms that could accommodate the large number of students that are commonly enrolled in a class today. But there was at least one that could. I got to class early on the first day of the semester. I was excited about the course I had chosen. It was linear algebra. Heck, before I started to college, this country boy thought that arithmetic was all there was. The mere fact of having discovered that arithmetic was only one of many branches of mathematics was a big accomplishment for me. Now there I was, the first student to be seated in the classroom to study something called linear algebra.

We students were instructed to come pick up the textbook for the course. The stack was huge. I don't remember the title, but the book was downright scary. I opened it up and immediately I was convinced I was in the wrong class. This textbook was one on Greek or something like it. There were discussions of vector spaces and linear mappings between those spaces. As if that was not confusing enough, the author then launched into a discussion of lines, planes, subspaces, and properties common to all vector spaces. I already knew about lines. I stood in a very long one when I matriculated. I also knew about planes. They were parked all over the airport where I worked. But these danged vector spaces were very bothersome. My first thought was, "I really don't think I'm going to like this course."

I do not remember how many students finally seated themselves by the time class started but the number was large as far as I was concerned. I guess the instructor felt the same way too because after he finished calling the roll, he announced, "There are too many students in this class and I'm going to get rid of half of you." He did too. The first rattle out of the box, he berated students who dared ask questions. I thought, "You are a crazy jerk, how can we learn without asking questions?" But, before I could get up nerve to challenge him, another student, a friend of mine, held up his hand and when acknowledged, asked, "Sir, what's a radicand?" Any first-year algebra student should know that. I tried hard to stifle a snicker, but it came out anyway as the instructor loudly exclaimed, "Holy horror to hell!"

Where Is the Runway?

I shall omit the name of the airline to protect the guilty although I doubt seriously if they really were guilty. This story begins in Anchorage, Alaska. I was a meteorologist working there in the nineteen-seventies and early eighties. Our work schedule rotated us through several areas of responsibility. We were responsible for putting out marine forecasts for weather and waves in the Gulf of Alaska, the high seas, as well as six thousand miles of coastline. Furthermore, in addition to public forecasts, we issued fire weather and aviation forecasts. The latter is the type of forecast on which this story will focus.

What is an aviation forecast? Well there are several forecast products designed for aviation interests. Basically, they include predictions of all types of atmospheric phenomenon that can affect flight such as clouds, precipitation, and visibility, as well as various flight hazards such as turbulence, icing, thunderstorms and wind shears. But lest I bore you with trivia, let me get right to it.

Because we provided products essential to aviation, we could take familiarization flights so that we could get feedback from flight crews as to how our forecasts were used. On such flights, we had permission to sit in the jump seat inside the cockpit after we briefed the crew on what to expect in the way of weather along the route. When the crew was not busy, we would ask them questions and offer input as to how they might better use what we gave them.

We would fly out of Anchorage and connect with another

airline in Seattle and then fly to Dallas. On this flight, my first such, I briefed the crew and seated myself in the jump seat. I was impressed at how many checks were made prior to takeoff.

As I watched the pre-flight procedures and listened to the crew interactions, suddenly an excited voice called out, "Pull up, pull up. You're too low. Pull up."

I jumped up and ran out of the cockpit screaming, "We're all gonna die. We're all gonna die."

Then I realized that we had not taken off yet. I am telling you, I was so embarrassed I wanted to die. Well, it could have been that way.

The engines revved up and down the runway we went, and I had a pilot's view of the concrete passing rapidly beneath us. While climbing to altitude, the flight engineer pushed the test button that engaged the warning system. "Pull up, pull up, you're too low. Pull up." Again, I dashed out of the cockpit looking for parachutes. As I looked around at the anguished faces of the passengers, I wondered if the FAA would let me take another FAM flight.

We were about two thirds of the way to Dallas when the pilot and second officer began checking their calculations. Their conversation was centered around the question, "Do we have enough fuel to make it to Dallas?" I was never sure whether they were seriously concerned or just wanted me to go looking for the parachutes again but about that time the flight attendant brought a meal to us and the pilot asked her to bring him some "Texas ice cream" (Tabasco sauce) and I knew all was well on Flight xxx.

On the return flight, a different crew came aboard. They seemed more serious. I never picked up on any conversations that I thought were designed to strike fear in my heart. We climbed to flight altitude and headed for Seattle. As we passed within view of Mt. St. Helens, the volcano erupted. Since it had been expected to do so, we were well away from any attending dangers. It was a sight to see, one I shall never forget.

The rest of the flight was uneventful until we began our approach to SeaTac airport. Seattle sits in a basin that often fills up with low clouds and visibility. On this day, as I remember, the area was experiencing a thick haze that hung low over the airport. As we descended, the crew began discussing whether anyone could yet see the runway. As we broke out of the haze, the runway was well to our left. There was a sudden scrambling to get the airplane lined up with the runway.

"I thought you had it," declared the pilot.

"No, I thought you did," came the second officer's reply.

Whether that little drama was another attempt to scare me or the real thing I suppose I will never know. No, I know I shall never know. The crew, considerably older than I was in my forties, have long since died of old age or something else. If they did it to scare me, it worked but I got even. I left a brown spot on the jump seat for the next meteorologist dumb enough to fly on a familiarization flight and hear the crew ask, "Where's the runway?"

God, Ice Cream, and Watermelons

Since I was a young boy, I have always liked ice cream. I remember the carefree days of my childhood when family and friends would gather on hot summer evenings to visit and wait until temperatures cooled down enough to go inside and sleep. Much of the time, Dad would pull a big block of store-bought ice from the old icebox. That was before we had a refrigerator. He would use an icepick to chip off chunks of ice and pack them around a canister that had been filled with an ice cream mixture made of milk, cream, sugar, eggs, and vanilla. Then we took turns cranking the handle on the old wooden slatted freezer. The ice was salted well to lower the freezing point of water. When the mixture began to solidify so that it was too hard for us kids to crank, the adults would finish the job. Then we all enjoyed the fruit of our efforts.

Right on the heels of my love of ice cream comes my love of watermelon. So good! Again, the watermelon, or ice cream family festival took place on the front porch. We would sit on the banisters and eat the watermelon as though we were playing a French harp. Sometimes the watermelons came from our field and sometimes from the store, but anyway you hack it, they were delicious. I think they were considerably larger and tastier back then. I still will not turn one down today though.

"But what does this have to do with God?" you might ask. Well, I am glad you asked. As I was eating a watermelon yesterday, I got to thinking about how long they may have been around. Were there watermelons in the Garden of Eden? I do not know. I am not quite that old, but I bet there were all sorts of melons there.

As I thought about Adam and Eve sharing a watermelon, I started wondering about the One who created melons in a way that I never had before. The bible teaches us that God is a Spirit. For our sake, He talks about keeping us by His strong right hand and promises us that He will walk with us. So naturally we think of God as having a head and a torso with arms and legs. He does not. At least the Father does not. God, the Son, Jesus, does though, for He came to Earth as a God Man and lived among us. His body is a glorified one now, as ours will be one day if we have trusted Jesus to save us. God does not eat watermelon, does He? No, I don't think so. After all, He is a Spirit.

Well, now, that thought really whetted my appetite for a deeper understanding of the Creator. For example, the bible teaches that God is all knowing, ever-present, and all powerful. The all-knowing part of Him means that He can enjoy giving us what He knows we will like. I am talking ice cream and watermelons here. I am talking cake. How does He do that? I can't answer that except to say, He just knows. You do not believe in God? Consider this. Originally, there were only two people on this large planet we call Earth. One can't say, "Oh the tasty things just evolved or were engineered in labs. No, they were not. The Garden of Eden was full of them. God thought them into existence. Yes, He did. I am so glad God knew that I would like ice cream and watermelons.

Crowley, A Town That Kept Us Alive

Some folks might have viewed Crowley, Texas, in the nineteen-sixties, as just a little mud puddle of a town where at any moment a stagecoach could round the bend or a saloon door might burst open with a fight spilling out into the street. It perfectly fit the bill as a town one might see in an old John Wayne movie. The population was only five hundred and eighty-six, not counting horses and cows. Yes-siree Bob, Crowley was just a tiny town in a very big state but what with Fort Worth being just minutes north, it was bound to not stay that way.

While it is true that Crowley, Texas might have been one of those "blink of an eye" towns where you would miss seeing it if you blinked your eyes as you drove through, it was no such kind of town to Rheta and me. *Au contraire mon ami.* To us, Crowley was a special town that indelibly stamped its importance on our minds. Crowley is a town that will forever bring smiles of gratitude with every thought of it.

During the nineteen-sixty-seven/sixty-eight school year at the University of Texas at Arlington we made a tough decision. I had been attending classes there while working full time as a weather observer in Fort, Worth since nineteen-sixty. If I continued to plug along at six hours a semester, it would take four or five more years to receive my degree. If I took leave without pay, it would be a rough row to hoe financially, but I could do it in a year. It also would mean that Rheta would be the one and only income earner. "Let's do it," we agreed. What a trooper she was! That is who she was. That is who she is.

At any rate, we began our one-year journey through the succeeding twelve austere months. They were good months though. In addition to my academic learning, living on one salary taught us that we were a team. We have been one ever since. One of the things we learned during that time was how to survive on ground beef. We called it hamburger meat. Oh, my goodness, we learned a gazillion ways to use ground beef. It was kind of like shrimp was for Bubba, Forrest Gump's friend. You know, hamburger spaghetti sauce, hamburger meat balls, hamburger helper, hamburger meatloaf, and that refrain just went on and on. We sure ate a lot of ground up cow.

Where did we get all that ground beef? We got it in Crowley, Texas of course at a place called The Crowley Meat Market if memory serves me. It was both a slaughter house and meat market. Back then you could drive into Crowley and stock up on good hamburger meat for thirty cents a pound. We bought so much hamburger meat there that, as I drove into town, the cows would flag me down and beg me to turn around and go back home empty handed. To us, it was a staple. For sure, by the time I graduated and went back to work, I would have welcomed a plate full of staples. At least they would have picked all the hamburger meat from between my teeth.

So, you can say what you will about Crowley, Texas but nothing will detract from the importance of the town to Rheta and me. Of course, it's not the same town now that it was back then, and we have not been there in many years. The once little "blink your eyes and it's gone town" has become a "cast your eyes upon me now and stay awhile town" of more than fifteen-thousand. Although the town has gotten much larger and perhaps the Crowley Meat Market is no longer there, the place will always be known to us as, "Crowley, a town that kept us alive."

Pride in the Red, White, and Blue

What does Old Glory mean to you? It means a lot to me. I can't ever forget my first time to don a United States Air Force uniform, extend my right arm fully out and parallel to the ground, bend it at the elbow, and then bring my perfectly aligned arm and hand smartly to my brow in a salute to the stars and stripes. I can't describe the flood of emotions I felt in that moment. *Just like my brother and brother in law*, I thought as I held my salute for bugle presentation of Taps as our flag was lowered in the evening light.

Of course, I was far from being just like my brother and brother in law. They had fought their way through Italy, France, and into Germany. They saw intense action in places like Sicily, Salerno, Anzio, and Monte Cassino. They earned stripes, badges, patches, and campaign medals. My sleeve was as smooth as a baby's behind. I had no stripes, not even one. I had not even earned a good conduct medal yet. I was an Airman Basic. But when I marched in parades and saluted the colors as we passed by, in my mind I could not have been prouder if I had been a highly decorated general.

I am both saddened and angered when I see news accounts of our flag, our symbol of freedom, being burned and stomped on. Don't they know that thousands of young men and women gave their lives to preserve the freedoms they enjoy? Surely, they understand that no amount of flag-burning will take away the deeply entrenched conviction that this country is worth dying for. I wish they would just one time lay down the torches, put on a United States military uniform, and salute the flag of the United States of America. Perhaps if they did, they too would feel the same pride in the red, white and blue.

Of Flashlights, Blankets, Books, and Far-Away Places

Rheta first told me of her love of books when we were dating. That was a few years ago. No, it was beyond a few. I suppose that lends even more meaning to what she told me. She said she was made to go to bed at a reasonable hour because she had school the next day. But she loved reading so much that she slept with a flashlight and a book. When lights went out, out came the flashlight and book. She would pull the blanket over her head like a tent and read until she fell asleep. During our life together, she has read many books as have I but as every parent knows, time and the busy work a day world takes a toll on such pleasures.

I kind of think that reading is more pleasurable for the very young, don't you? As we grow older, reading becomes more for acquiring information. After a point, the books we read bring back memories rather than pointing to possible new adventures. In our childhood years, books freed us from the constraints of the time. We could not take a train, plane, or boat and go visit distant lands, but we could go there via books and magazines. Books and magazines became our transportation.

Being transported by the pages of books enabled us to see places we might never see although some we have seen. We could also travel to the imaginary land of Teeny Weenies, comic book characters only two inches tall. Or, we could cross the same bridge that the three Billy Goats Gruff passed over. Remember? The big Billy took care of the troll that lived beneath the bridge and threatened the little Billy Goat Gruff. Because of books and magazines, many places have been visited where our feet have not yet trod but we shall always remember the ones we read as kids in the days of flashlights, blankets, books, and faraway places.

It Was A Mighty Good Roast

I grew up in the little southeastern Oklahoma community of Panola. For me, a cherished memory was of the railroad track that ran through the small community. The school bus would stop, whether a train was coming or not, and let one of the students off to signal that it was safe to cross. I always loved it when I was called on to perform such a "highly responsible" task. It made me feel so important (tongue in cheek). It was kind of like being assigned to "dust" erasers.

The afternoon train was the one that brought our mail. The conductor would lean out and snag a pole that would accept the bag of mail he carried. Then the train, pulled by a coal-fired steam engine, would rumble on down the track and out of sight. It was just such a train that killed ole Snip, my blaze faced sorrel that got out of our pasture and ventured up onto the tracks.

It was also such a train that hit and killed a neighbor's cow. The neighbor told our vocational agriculture teacher to bring his class and salvage the meat. We could divide up the meat. I loved our agriculture class. We frequently got out of class to go on field trips to help people with their animal's medical needs. We branded, vaccinated, and castrated many animals. Our teacher was a smart guy. He should have been a vet. He made short work of the butchering job and we learned a lot from watching him. I lucked out and was given a roast from the shoulder of the cow. I was sorry for the neighbor's loss and for the poor old cow, but I must say that it was a mighty good roast.

Walk in the Light

One has only to awaken to a new day to see that we live in a dark world. The news continuously informs us of murders, corruption in our government, thefts, immorality in heretofore respected individuals, and on and on. When will it stop? When will there be peace? Never on this earth will it end. That is why we must find the light and walk in it. Is it possible to do that? Yes, it is. But it is also extremely difficult. We struggle to keep our fleshly desires from overcoming us. We falter. We fail. We get up and keep walking where the shining light directs. Then it becomes easier.

But where is that light? The light is in Christ Jesus and He is in all who believe upon His Holy name. Think of it as if you were a small child walking with your parents on a very dark night. Mom or Dad shined a flashlight on the road. If you got out of the path of that light, you immediately became frightened by the dark and hastily retreated into the light where you felt safe. So, when the world frightens you, ask yourself whether you might be paying too much attention to the world's distractions. The Holy Bible is God's flashlight to keep us from wandering into the darkness. He shines the flashlight on the right road. All we must do is, "walk in the light."

Khan, A Dog Worth Remembering

He was solid black. Even his tongue was black. He was half chow. The other half was black Lab. He had the longer hair and curly tail of a chow. His temperament was certainly that of the chow. He was given to Rheta as a pup by a doctor with whom she worked and right off the bat he established his self-perceived dominance by biting the doctor's hand as he pulled him from a cave beneath a giant spruce tree. Perhaps, a subconscious thought surfaced of Samuel Taylor Coleridge's Mongol dream emperor Kubla Khan. Whatever the reason, we christened him Khan. Oh, how right we were in doing so for indeed, Khan took no prisoners.

We took this black-tongue warrior devil home and he became part of the family. He immediately owned our young son and daughter. The only reason he did not own Rheta was because she was quicker on the draw in the bonding process. Each morning Khan would lie on her lap while she had her coffee before getting dressed to go to work.

Discipline for Khan was performed by taking hold of the loose skin of his cheeks and staring him in the eyes. He hated that. Once, after receiving his scolding from me, he promptly bit my hand as I released my grip. From then on, I was faster doing it.

This wonder dog was very protective of the family, especially Rheta. Once, when I came home a little past midnight after working the swing shift, Khan was lying on our bed while Rheta slept. He decided he was not going to let me get in bed. When I reached for the cover he snapped at me. I had to awaken Rheta, so she could intercede for me. Yes indeed. I knew then we had given this royal Turk the appropriate name.

Khan never thought that he needed permission from anyone to do anything. He sure had a mind of his own. One Thanksgiving, Rheta had prepared our meal but was sick and so after we ate, she left the roasted turkey on the table for us to put away. She went back to bed and we followed to tuck her in and make sure she did not need anything. When I returned to the kitchen a few minutes later, the platter held the bony carcass of our turkey. The thief was nowhere to be found. Alas, it took us some time to find him. I think we must have passed by that big black round ball in the corner of the room a hundred times before realizing it was him.

After leaving Alaska, we moved back to Springtown, Texas where we had a house built on our acreage. Carrie was in middle school and loved ducks so when I saw some baby ducks for sale at the feed store I bought two and brought them home. Ducks love water so I put them into a tub and went into the house. I was not aware that baby ducks should not be left in water too long since they have no feathers and thus no protective oil. When I went back, one had died, and the other was suffering from severe hypothermia. We did manage to save it by wrapping it in a warm towel. I buried the duckling under a giant Live Oak tree in our yard. Khan was observing me the whole time. I got up to leave and, after taking a few steps, looked back. Khan had wasted no time. He had dug up the baby duck. I watched as he gently took he tiny creature in his mouth. That rascal dog was about to eat that duck, I feared. But, my curiosity was stronger than my thought of scolding him, so I continued to watch. He carefully laid the baby duck on the ground next to the hole and walked away. He had saved the baby duck, so he thought.

Rheta was not working then and would often go visit our friends Bill and Clara Nell. It was only a short walk if she cut across our neighbor's pasture. The only problem was, our neighbor had a big black angus bull. He did not like Khan. But, the bull was across the pasture so Rheta figured she could walk the short distance without disturbing him. Wrong! She had not figured Khan's protective nature into the mix and he promptly ran out toward that black mass of bovine flesh, faithful to his calling as Rheta's body guard. The bull did not like that one bit and, with head lowered and nostrils emitting steam, came at woman and dog like a locomotive with no brakes. The fence was closer than the bull so they both made it to safety.

Few dogs have pleased us as much as he did. A large oak tree now shades the grave of Khan, a dog worth remembering.

Thanksgiving Day
Two-Thousand-Seventeen

I do not know if it is an age-related thing or something else, but I think I have thought more about being thankful this Thanksgiving season than all the others. At any rate, when I think or write down what I am thankful for, I am compelled, by the goodness of God, to include Him first. He gave me eternal life by the blood of His only begotten son. So, I thank Him first. And, I thank all the great men He used as they wrote what He dictated as scripture. I am grateful for His goodness and mercies which are "new every morning" and for the eighty years He has given me. I am also thankful for the family He gave me, especially for Rheta and my children and grandsons.

I thank God for the nation in which I live albeit wounded as it is. We still have the Constitution. We still have religious freedoms. We still can speak our mind freely, although sometimes that freedom seems to be slipping away. Consider the cost of the freedoms we enjoy. Thanks to Wikipedia, we can peruse the numbers. The task of listing all the links that lead to war statistics is far too daunting to attempt. So, I have listed a main link here in order that those interested can refer to the numbers.

https://en.wikipedia.org/wiki/United_States_military_casualties_of_war

Notice the total number of casualties. Starting with the American Revolutionary War, and going forward to the battlefields of today, the column "Total Casualties" tells it all. Well, no, the numbers really do not come close to telling the whole story. For each casualty there is a story that goes untold.

When reading the list, we can't begin to appreciate the story. In the record of total casualties, we can't see the flood of tears streaming down the faces of mothers as they hear the words, "your child was killed in action." Nor does it announce the anguish of the young wife or husband as they grieve the loss of their spouse and contemplate raising their children alone. The numbers do not speak to the pain of the soldier lying on the battle field writhing in agony while waiting for a medic or to the many fighting men and women lost at sea. A column of numbers can't address the abject fear in the heart of the young man or woman charging full into an oncoming hoard of warriors. Numbers are so pitifully inadequate to express reality in matters like these.

When reflecting upon our freedoms and how they are preserved, we can't, indeed, we must not fail to mention another segment of our society, as important as our military. Some might not see that first responders are essential to our freedoms, but they are. From the doctors, nurses, and emergency medical technicians who, by their skills, prevent death tolls from rising during natural disasters and terrorist attacks to the men and women in blue who speed toward trouble in the night, to the fire fighters who rescue people from burning buildings, we remain alive and free, in no small part because of them.

As we celebrate Thanksgiving Day this year, let us remember the price that so many have paid for so many freedoms we enjoy and pray for special blessings upon those who still live and strive to keep us free. May we consider, as we sit back in the comfort of our homes after feasting upon traditional Thanksgiving dishes, that not everyone is free. Not everyone has a warm house and comfortable bed. Not everyone will end the day with full stomachs. So, as we give thanks to God for our freedom and for our blessings, I hope we will put forth a petition to the almighty that He will extend great mercy and grace to those less fortunate than ourselves on this Thanksgiving Day two-thousand-seventeen.

Henry was a Prankster, Too

His house was about a quarter mile off the main road on the south side of the railroad track that still runs through Panola, Oklahoma. The old Henry Wilson place is gone. A new family lives there now. A new house stands where the old red asbestos sided house used to be. I doubt that you will find Road Island Red chickens scratching for bugs there now. The old gray goose that used to scare me when we visited there has long since gone to wherever it is that geese go when they die, as have the Wilsons.

In my mind, all is fixed forever just the way things were when I was a kid. I still run around in the yard chasing that old goose after I found out that I scared him more than he frightened me. I feel the warmth of the old propane fired stove as I back up to it to get warm and then must move away when it starts burning my legs. Mrs. Wilson's quilting frames hang from the ceiling and she is working on another beautiful quilt. I smell Maxwell House coffee and can hear the gravelly voice of Henry as he tells another story. I hear his familiar laugh. But distractions sweep me away from that house and all its memories, so I will just tell a story that happened there.

Henry and some of his buddies went duck hunting early one cold winter morning. He had an old double-barreled shotgun that always rested on his shoulder on duck hunting trips. Normally, if there were ducks to be had, Henry would have himself one or two. On this day, though, the ducks just did not cooperate, and Henry was forced to go home emptied handed. He walked to his car and was about to put "Ole Betsy" in the car when a bunch of crows few over. Reflexively, Henry swung the shotgun to his shoulder, took aim and fired. Down came a crow. Right then and there, he skinned and gutted the black feathered foul and put it in his jacket pocket. When he got home, he promptly took the dressed bird into the kitchen where his wife was preparing dinner (lunch to most people these days), He handed the crow to her and said he sure would like it if she included it in the menu. Mrs. Wilson look at it rather askance and commented on its appearance.

"Aw it just a skinny ole wood duck," he replied.

"It might be a little tough."

Feeling duty bound, Mrs. Wilson served the "skinny ole wood duck" on its own platter. She tasted it, but told Henry she was not very hungry. I guess he was not either because he did not even taste it. But this little trick proved that, indeed, Henry was a prankster too.

Nightmare at Midday

It is eleven-fifteen, Thanksgiving Day, two-thousand-seventeen, and I am peeling potatoes. I am just taking off those skins like crazy, dumb, fat (well, I have lost down to one eighty-eight so not so fat) and happy until I glance down and notice that I am wearing a shirt that is Army green. Yikes! Memories suddenly surfaced from my Air Force days of long ago.

At four o'clock in the morning the drill Sargent would flip on the lights, yell out a few demeaning terms of endearment and announce that the chow hall awaited. It was KP (kitchen police) day. We all knew what that meant. For hours we would be peeling potatoes and scrubbing pots and pans.

As my thoughts were interrupted by conversation with Rheta, I forgot about those days. Well, at least for a fleeting moment I did. Then the nightmare returned. Potato peeling? Army green shirt? Oh my! Horrors! The evidence is indisputable. I am in the army now!

Let There Be Light

"In the beginning God created the heaven and the earth. And the Earth was without form, and void; and darkness was upon the face of the deep. And the spirit of God moved upon the face of the waters. And God said, let there be light: and there was light."

What does our expanding universe have in common with trains and red shifts? First, with great haste, I confess an extremely shallow knowledge of the intricate parts of Hubble's Law. So, what you get here is a discussion of it from a country boy who minored in physics.

Interestingly, both Genesis and scientists for the most part, agree that there was a beginning before which was total darkness. But agreement ends there. The bible states that God spoke light into existence while scientists speak of something called the Big Bang (no, not the television program).

A few years ago, I read an article in Scientific American that addressed the Big Bang. The sum and substance of the article was that scientists were able, by various equations and methods, to trace the universe back to an age of an extremely small fraction of a second. In scientific notation, that age was ten to minus thirty-fourth power. Furthermore, the author of the article admitted that he had no explanation of what triggered the Big Bang and events prior to it.

Oh, and what about trains and red shifts? Well, both are Doppler effects that can determine velocities as objects move relative to each other. Have you ever stood at a railroad track and listened to the sound of an approaching train? The sound pitch (frequency) increased with the train's approach. As the train passed, the pitch decreased. Even if you had been blindfolded you could have told whether the train was moving toward you or away. While it is true I have talked about sound waves here, the same is true for light waves. Scientists have observed a red shift in observations of stars. According to Doppler, the red shift can only happen if the star is moving away from Earth. If the star was moving toward the earth, then the wavelengths of light would be "stacking up" and becoming shorter or of greater frequency thereby resulting in a blue shift.

So, our universe is expanding. Scientists can't say what happened before that tiny, tiny fraction of a second after the Big Bang. Nor can they say what triggered it. I can. God said, "let there be light."

No Siree Bob, Rudolph Ain't No Sexist

Now isn't that just special? The political correctness police have gone way too far in their zeal to turn us into a generation of genderless gyms of gutless, gloomy, glassy-eyed grownups. For the past decade or more they have been attacking Christmas by trying to besmudge the one thing that all small children the world over equate with Christmas, Santa and his reindeers. They said, for years that Rudolph the red nosed reindeer is misogynistic. You know that all the reindeer that pull Santa's sled are male. Everyone knows that Rudolph is the leader. So, he needs to go. That's right, ole Rudi needs to go, as does the foolishness of celebrating Christmas. Oh, and do not even mention Jesus. That sure will incite rage among that crowd.

We might as well shoot ole Rudi and put him on the dinner table. He is done. Never mind the countless nights when, as a child, I laid awake to see if I could hear the faint sound of Santa's sleigh. After all, that jolly old St. Nick had a bag full of toys. One of them was for me. And, I loved his reindeer too. Would I ever be able to look out the window and see the glow of Rudolph's nose? I always hoped I would but then I would drift off to sleep. The next morning, I would only see evidence he had been there, namely the toy he left for me and missing cookies and milk on the table. Was I concerned that there were no girl reindeer? Was any little girl concerned about such a thing? Little girls were just as excited as I was. They too were anxiously awaiting the sound of jingling bells that would herald the arrival of a bag that contained a toy for them. They did not care about the gender of those majestic creatures.

I wish we could still have some of those days today. Let Santa hook his sled to eight female reindeer. I am okay with that. In fact, everyone knows that girls run faster than boys. (Oh my! Such a sexist comment). Seriously, I am okay with however the commercial aspect of Christmas is presented so long as Jesus Christ remains as the main force behind the celebration. After all, Christmas is about giving, and Jesus gave all to redeem mankind from sin. He freely gives eternal life to all who accept Him as Lord and Savior. So, go ahead and create eight other reindeer and make them all females. Put ole Rudy in a nice warm stable and feed him the best grain you can find. Just do not ever call him a sexist because, no siree Bob, Rudolph ain't no sexist.

The Longest Time on Death Row

Who holds the record for the longest time on death row? Most people will look it up and answer, "Gary Alvord of Florida." Gary died during his thirty-ninth year on Florida's death row.

I would answer differently though. My answer to that question is Jesus Christ. Does that sound strange? Perhaps to some it does. However, if one believes the Bible, then it does not, because the Bible teaches that before the foundation of the world was laid, Jesus had a mission. He was bound, by agreement with the Father, to one day go to Earth and die a death of crucifixion to pay the penalty of sin of all who would accept Him as Lord and Savior. 1 Peter 19-21 speaks of this plan.

What would this mean to the son of God? Certainly, it meant that He understood what was to happen to Him. He, co-equal to the Father, would voluntarily give up his position in Heaven and be born on Earth, live a perfect sinless life and then lay down that life by dying on a cross to redeem lost mankind. Jesus understood that "it is impossible for the blood of bulls and goats to take away the sins of the world." Hebrews 10:4. A perfect sacrifice was required. He would be that offering. Hebrews 10:11-15.

Jesus also understood that His greatest agony would not be the physical pain He would endure but, rather, it would be a suffering of a different kind. The torture that would cause Him to sweat drops of blood (some think hematohidrosis) while praying in the garden of Gethsemane as His crucifixion drew near would begin his torment. The piercing of His side with a sword and the nails that ripped the ligaments of His hands and feet as He hung on the cross did not cause His greatest pain, nor did the crown of thorns that caused blood to run down into is eyes.

The prophet Isaiah said, "But your iniquities have separated between you and your God, and your sins have hidden His face from you, that He will not hear." Isaiah 59:1-2.

Think about that. His emotional pain was so great that He cried out, "My God, my God, why have you forsaken me?" Matthew 27:46.

Why the passion of the cross? It was the only remedy for our sin condition. The sin of Adam and Eve in the Garden of Eden put a curse on all of mankind. Every person is born with that original sin condition. It guarantees that we will perish. But we do have a remedy. To atone for our sin, "…God so loved the world that He gave His only begotten Son that whoever believes on His name shall not perish will have eternal life. John 3:16. It is a gift. We do not have to earn our way into Heaven. We only must believe on His name. If you have not placed your trust in Jesus, there is no better time than now to do so. After all, this Holy season of Christmas is about Him. It is for you. It is all about the One who spent the longest time on death row.

In A House of Eleven, Old Hens Did Not Have Long to Live

I have always loved chickens. They were, of course, a necessary part of farm life in rural Oklahoma when I was growing up. Momma would order baby chicks from Sears and Roebuck's catalog whenever her flock numbers got low. We mostly had Plymouth rocks because they were pretty good laying hens and excellent meat chickens.

Our chickens always ranged free except when dusk came, and they headed for the chicken house to be closed in for the night. Dad, being a carpenter, had constructed an excellent house for them. There was a long, wide opening in the front and a stretch of chicken wire across it for ventilation. A hinged door could be lowered and latched when the weather was cold and wet.

The front yard was fenced off so that the chickens could not destroy the grass. Those suckers can go through a grassy yard like a swarm of locusts and all that is left after a few days is dirt. The flock seemed to favor the back yard so consequently there was nothing but bare ground there and one had to take care where they stepped. I always hated when I stepped in chicken poop because most of the time I went barefooted in summer.

As all farm kids do, we only gave names to a few animals. We only invested our affection in ones we knew would not be eaten. For some reason, we did not have names for chickens or, if we did, I do not remember it except for one chicken. Even now that one stands out in my mind. Maybe I remember because it was the only one my sister Nadine and I named. Or, perhaps it was because of the condition of the chick. When the new order of baby chicks arrived in the mail and we brought them home, we found that one was born with a deformed leg.

Nadine and I took that little fluffy ball to raise. We named it Crippled Feedbriar. How we came up with that appellation I don't know. But I remember being followed around by that little rascal as she would flop along behind my sister and me.

Momma kept close tabs on her chicken flock and knew which hens had outlived their productive years. Once they slacked off on laying, it would not be long until they made a trip to the chopping block.

Momma had fashioned a chicken grabber out of a clothes hanger. We would throw corn on the ground and then carefully slip the grabbing end of the wire against the appointed hen's foot and yank. Usually, we would catch two. That is what it took to feed our large family. I was too small to wring the old hen's neck, so I used a hatchet to chop off her head. Momma would take hold of the neck and swing the hen in a small circle by her side three or four times and off came the head. I was always fascinated that even with her head gone, that ole hen would run around all over the yard before it finally collapsed. You might say she was running around like a chicken with its head cut off.

At that point, the soon to be chicken dinner was dipped in scalding water briefly to loosen the feathers for plucking. I would stand by the kitchen sink and watch as Momma butchered the hen. The egg duct of younger hens would have one egg ready to lay, followed by a train of several other eggs of decreasing size.

Farm life was fun for kids, hard work for adults, and just plain healthy for everyone. We ate from a large garden, had milk and butter from our cows, bacon and hams from our pigs, and eggs from our hens. Well, we had eggs from our hens until they stopped laying. When that happened in a house of eleven, old hens did not have long to live.

Past and Future

I wrote the following poem when I was in college long ago. I was taking a course in English and the topic was personification.

Past and Future

Past held Future in his lap
But Future could not be still.

Rending the chains that bound him there
He imposed upon Past his will.

So, there Past sits and there he laughs
And watches as the moments pass.

Past content because he knows
At whose expense, he grows and grows and grows.

The Fire

I wrote this poem upon learning that the old house where I was born had burned to the ground.

The Fire

What did you take with you
When you engulfed that which was mine?
My memories were there, part of my heritage.
Surely you knew. I told you.

You came without my knowledge,
And burned it to the ground,
The house, the room where I knelt at night
My prayers to say.

How can I remember now
All those things about one's birthplace
That should be remembered?
I could have though
When the house stood.

Even the road that led to the house is washed out
And overgrown with bushes and weeds,
So that my memories must remain
Hidden there among the ashes.

DONALD MANKIN

What did you take with you
When you engulfed that which was mine?

She Could Have Been from Royal Stock

I knew a lady once
She was unpretentious but had no reason to be.
And she was always friendly.
She could have been from royal stock.

I knew a lady once.
When I hurt myself, she cried inwardly.
But outside she managed calm
Like a teaching fellow manages his stipend.
She could have been from royal stock.

I knew a lady once.
She gave me a calf from her only cow.
I kept it until it had a calf.
Then I sold them both
And bought things for myself. She was happy.
She could have been from royal stock.

I knew a lady once.
She could touch a wilting flower
And it would spring up
More alive than a thousand fresh daisies.
She could have been from royal stock.

Tornado

The night was silent, so silent, forever it seemed;
And many waited for the waking, for the breaking, for the sound.

When 'twas thought to be eternal
Then it came, so loud it screamed.

And dreadful moments turned to hours;
As they waited to be found.

The Taste of Texas

The sights and smells are here to stay,
Of meadows green and fresh mown hay.
I am told there is no other way
To savor Texas' flavor strong,
And live a life of joy so long.

A Bridge

A bridge that spans a quiet lake
Is just the place where one can take
A troubled soul and there, unload
All burdens for his inward sake.

A bridge that hangs o'er waters still
Is where one can impose his will,
Upon those things that sting his soul
And never let him drink his fill.

A bridge of rope, or wood, or steel
It matters not if we can feel,
The solace that we're seeking there
And find the means our hearts to heal.

Winding Road

A winding road, dips down into,
The valleys of my mind.
And brings me home, old fields to roam,
A sweet and final time.

Jody's Owl

You sit staring at me with eyes that are free,
With eyes quite as piercing as hers.
But why must you hide in the forest from me,
Among the Aspens and Firs?

I told my Jody you wouldn't come here.
Said I, "he's far too wise.
Tis hunting season and I fear,
He'll be shot to the ground if he flies."

So, we searched the day through until it was late,
And then walked home feeling sad.
But as we drew near to her front gate.
You were perched there looking full glad.

Moments

Night slips by silently, softly,
Like September's leaves. I complained to Time,
But to no avail.

I told him it was the
Wednesday of my life,
And that I needed another
Unhurried, undisturbed moment
With you.

I said I wanted the moment
To savor the delicate, fluttering
Fragrance of your
Perfume.

I added that my moments with you
Were like the reflections
From dozens of dewdrops and
Could never be captured again.
But Time only laughed
And promised to send Night again,
Tomorrow

Friends

I wrote this poem, my first, while taking a literature course in college. My father and a very close friend of his would often sit on our front porch and play checkers. They were both up in years and his friend had Parkinson's disease. I called the poem "Friends."

Friends

His eyes were sad,
His head bowed low,
Expressing his desire to go.

I could not see his inward thought,
Nor tell which battles he had fought

But Time had tested his free will
And more than ever plagued him still.

The sun is low, it's getting late,
A rotten porch and rusty gate

Can never say "Come back my friend
So, I shall not come here again.

Candles, Kerosene Lamps, and Electricity

Our kids did not know too much about the early life of their parents. They didn't know that we studied by coal oil lamps. Perhaps this brief discussion will lend some measure of cause for them to better appreciate life's comforts today.

We had no electricity until I was eleven years old. So, we kept candles and matches at hand on the dressers in our bedrooms. It is true that we had flashlights, but the candles provided extra insurance should it be that we reached for a flashlight and found the batteries had run down. We had no indoor plumbing, so we made frequent use of flashlights when we had to pay a visit to "a man about a dog" (our way of saying, "I have to go to the outhouse").

Two or three kerosene lamps were present in the living room, dining room, and kitchen. Most of the lamps were just plain vanilla lamps. They had an oblate shaped glass globe that tapered a bit at the bottom to fit onto the base of the lamp and the diameter decreased at the top. A cloth wick extended from the burner down into the bowl of kerosene that comprised the base of the lamp. A control much like the winding stem of a watch facilitated the increasing or decreasing of the flame for light. When it was time for lights out, Momma hovered over the globe and gave a quick puff of air through her lips to extinguish the flame.

We had one "Aladdin" lamp which had a more beautiful base and taller globe. The taller globe gave a brighter light, and it was the lamp we used for doing our homework.

Daily cleaning of the lamp globes was essential, as was the trimming of the wick That job fell to us kids. We also kept a lamp in the root cellar where we took refuge from storms.

When I was eleven, we were excited to see a big REA (Rural Electrification Administration) truck coming down our private road. Electrification of rural Oklahoma was life-changing. It meant that soon there would be light in every room. And, we would not have to clean lamp globes nearly as often. It meant that dad would have the benefit of a skill saw rather than the hand saw he used to build our house.

I still remember the big auger on the truck as it drilled a hole for the pole. The pole was set. The wire that would soon carry the magic current was strung. The fuse box was set. The handle on the box was pushed up to complete a circuit and, suddenly, there was light. I remember those days with fondness, the days of candles, kerosene lamps, and electricity.

Country Boys and Girls Know About Things They Do Not Even Know They Know

Say what? That is right. They do. Take for example, a very cold and windy January morning. They know where they will find their ole milk cow Betsy and her baby. They will be huddled against each other in their stall. If there is no stall, then they will be lying together on the south side of a structure. A large herd of cattle will lie down touching each other in a large circle. Horses out in the pasture will stand with their rumps to the wind. On very hot days, pigs will wallow in mud, dogs will pant, and horses will sweat. These behaviors result in a higher level of comfort for the animal. It was not until I studied agricultural meteorology that I made the connection as to the whys of such animal behaviors.

Animal behavior with respect to weather will reveal several applications we did not even consider when we studied physics in high school or perhaps college. First, there is something called preferendum. My spellchecker flagged the word, so I went to the dictionary but was unable to find it at all. Nor, was I able to find my notes from my biometeorology lessons. However, the word preferendum simply means a place of optimum comfort. During extremes of weather, animals will always seek their preferendum.

So, thinking back to ole Betsy and her calf and connecting that huddling behavior to what we learned about heat transfer in high school physics, we recognize the process as heat transfer by conduction. In this case it is a temperature gradient from the warmer cow (she has greater size and thus more heat) to the colder body of the calf. In the circle of cows lying huddled together we see the process again.

What about the horse that stands with its rump to the wind on a cold windy day? Do a google search and you will get all sorts of crazy answers like, "it doesn't like wind blowing in its face," but the real reason has to do with how equine core temperature is conserved. In the average sized horse, about ten gallons of blood spends more time in the front end of the horse where the heart is located and where the blood is closer to the surface than in the rump. So, by standing rump to wind, the horse is instinctively protecting its core temperature from a dangerous cool down.

So where is the pig most comfortable on a hot summer day? A mud puddle, of course. Every country boy or girl is accustomed to seeing mud caked on the pig's body in the summer time. Wallowing in mud serves two protective functions. For one thing, it protects the skin from sunburn, and secondly, evaporation cools down the pig's body to keep its core temperature from rising too high. Shady areas also are more comfortable.

Comfort wise, for animals in winter, anywhere they can get out of the wind is where you will find them. It may be the south side of structures or erosion ditches. Regardless, you will know one thing about them. Where they are found on a cold day will be the location that affords the greatest comfort.

There is so much more involved here like how body hair stands up when the weather is very cold. More air can be trapped between the thick hairs and air is a very good insulator. That is why we pile on more quilts at night when we sleep in a cold room. Air is trapped between the quilts. I love sleeping in a cold room. I say, the colder the better.

Animal preferendum is an interesting topic that mere observations can't do justice to. If you are a nerd like me, let me encourage you to research the subject but be careful using the internet. It can lie you know. There are some really goofy explanations out there. If all else fails, ask a country kid because country boys and girls know about things they do not even know they know about.

One, Two, and Many

The Warlpiri, an indigenous people of Australia, once upon a time long ago felt no need for an elaborate counting system. Most could count to two in their language and a few managed to make it to four. Beyond that, and generally, they counted, "one, two, and many."

The Apostle James told us that, "Every good thing given and every perfect gift is from above, coming down from the Father of lights, with whom there is no variation or shifting shadow." James 1:17. As I reflect upon all the good things I have received at the hand of my Heavenly father, I feel like the Warlpiri people must have felt, a bit frustrated.

Oh, it is not that we have an inadequate counting system. Instead, in my case, it is my memory that fails me. Surely, we all remember and can number the notable "good things," but are not ALL good things noteworthy? Yes, I think so. For example, when I was a kid my saddle girth broke while running my horse. I landed in a ditch. I was not hurt. That was a good thing.

Or, how about the many summers when I ran around barefooted all over three hundred and fifty acres? There were snakes, broken glass, rusty nails, and other hazards such as sheer drops of fifty feet or more on the bluffs where I often played alone. Did I ever give a thought as to those things being "good things" or "blessings"? No, probably I did not. But, you see my point. They were good things. They were from the "Father of Lights."

Those "good things" require some measure of reflection. What does not however are the good things like a wife who, indeed, is the wife of my youth and the children she bore to me and the grandchildren my children produced. Those are the best things. To number all blessings would demand more time than I have and probably find me saying, "one, two, and many." Praise the Father of Lights for the "good things."

Prom Dresses in the Sky

It was Sunday evening, well past dark since it was winter time in Anchorage, Alaska. We had managed to navigate the snow-covered road that wound its way up a hill and out of Bear Valley where we had enjoyed a day visiting the McLaughlins. I can still hear Geraldine singing "It is well with my soul." About half way down the mountain side (the Chugach range) it happened. What a startling sight! Was it the second coming of Christ? Was it a Damascus road experience like Paul had? No, but it was majestic. We were looking at the most intense display of the aurora borealis (northern lights) that we had ever seen.

We were accustomed to seeing these magnificent phenomena many times but most of the time they were visible near the horizon out our bedroom window. These, however, were high in the sky, almost overhead. Never had I seen the lights so intense. Certainly, I had never heard them. Scientists used to believe that they occur too high up in the atmosphere to be heard. Now we know that such is not the case.

I do not know when it was first discovered that indeed they do talk (my words, not theirs). We personally experienced the atmospheric noise as we got out of the car and stood looking up until the cold night air drove us back into the warmth of our vehicle. Perhaps it was not the cold that caused us to take refuge. Maybe our subconscious feared that one of the great dancing curtains of light would come unhinged and fall, smothering us in its brilliant folds.

I heard an eerie swishing sound. I say eerie because it

sounded much like chiffon prom dresses on a dance floor. That's it! That is what it was! What appeared to be curtains were not that at all. We were really looking at prom dresses dancing their way across the Van Allen radiation belt. The "swish, swish" was followed by the crackling of a campfire but we were not camping. There was no time for camping anyway. We had to leave that wonderful performance on God's stage. The evening worship service would be starting at Sunset Hills Baptist Church soon. We loved that little church so much. Neil Thompson was the pastor there. I sometimes post his inspirational poems and prose on my Facebook page.

After many years we recently reconnected with the Thompsons. And, for the past several years we have been Facebook friends with Art and Barbara Braendel and most recently with Geraldine McLaughlin and Shirley Miebs, all of whom were members at Sunset Hills.

I wish I could reconnect with the aurora borealis but, alas, I live too far south. On very rare occasions the lights can be seen in mid latitudes but the odds against that are extreme and Bear Valley is so far away. So, I suppose I will just have to dream of good friends, good sermons, good music, and of course, prom dresses in the sky.

Henry and The Talking Deer that Threw Rocks

Once, in a previous writing, I alluded to this amazing animal though I have no personal experience of having seen him, so I can't say if he really existed. I only have Henry's word for it. Was it fact or fiction? Does it matter? Whatever way you slice it this deer is no longer alive. If he was real, he either was killed by a hunter or has died of old age long ago.

On our three-day fishing trips with my dad, Uncle Press, and Henry I never tired of hearing his stories. I would beg Henry to tell the story about the talking deer. He was always most accommodating when it came to storytelling. I think he knew that the talking deer story was one of my favorites so each time he told it he would tell it better than the last time and I rewarded him with great belly laughs. The more I laughed, the better the story got.

The tale goes something like this. Henry was up in years and not in the best of health so when a group of men went deer hunting they would let Henry off at his favorite stand only a few yards removed from the road. The rest would go on a bit farther and then stop and hunt on foot. One day, Henry was on his stand early and still full from a good breakfast of bacon, eggs, and biscuits. He had arisen early to make a fire and put a bucket of water on to boil for coffee so, on the stand, he drifted off to sleep.

The road had clusters of loose rocks here and there. Some were small, some were large, and some in between. The noise of rolling rocks jarred Henry to an awakened state and he saw the cause of the rockslide. According to him it was the biggest buck deer he had ever seen.

"Boy if I could put that rack over my fireplace I would be in hog heaven," Henry thought, as he reached for the old forty-four forty he hunted with.

The only problem was, Henry got so excited that he gave out an expletive and rattled his bed of leaves.

"Hey, put that thing down," ordered the buck as he flared his nostrils and pawed the ground with head lowered.

"What do you think you are doing?"

"Oh, I'm just messing with you," replied Henry.

The buck must not have been too sure about the truth in that because he hesitated just long enough for Henry to put his gun sights on him. He fired off a shot, but in his excitement, Henry missed by a country mile. "Hey, what are you doing you crazy ole fool? "By then the buck was looking mighty serious as ole Henry tried all the more to load another round in the chamber. His gun had jammed.

The ole buck had figured out that he was in danger of being meat on Henry's table by evening unless he did something fast. So, he angled himself at about one hundred-eighty degrees and started pawing the ground in the rock pile. Henry, realizing the seriousness of the situation, frantically worked to chamber another bullet. All the time the buck was snorting and pawing the ground. Then Henry became enlightened. That ole sixteen-pointer was not just pawing the ground because he was mad. That crazy fool deer was throwing rocks at him.

This back and forth went on for some time and then, for some reason the buck stopped throwing rocks and Henry stopped trying to get another shell in the barrel. They stood for what seemed like an eternity to Henry as he contemplated the power of his enemy. Then after sizing things up, the deer made the first gesture. Snorting out a few choice words, he turned and walked away muttering "Go on home you ole fart. You're not worth killing anyway."

The Five-Year-Old Rancher

He was only five-years-old when I gave my son, Steede, his first ranch-hand position. I was working as a meteorologist for the National Weather Service in Fort Worth and lived in Forrest Hill. Being the country boy I was, I had long been looking for acreage that we could afford to buy and still be close enough to commute to downtown Fort Worth. Rheta was working for a doctor whose office was in the Westchester House on Summit street. We had lived there before Steede, our first child, was born.

Our apartment was a one-bedroom efficiency with a tiny little kitchen. Living there was quite an experience. If you got something from the cook stove, you better not have the refrigerator door open or you would freeze your derriere. And, if you got something from the refrigerator, you better not have the stove burners on. So, you can see we had to find living quarters with room for a new addition to the family.

Rheta has always been good at finding us a place to live. I have never disagreed about the places she chose. On at least two occasions, she drove around in the neighborhood where she wanted to live and looked for houses that were for sale. In Forrest Hill, she drove around a neighborhood looking for houses for sale. She spotted a house that appeared to be vacant.

The grass was badly in need of a lawn mower and there were no blinds on windows. So, she got of the car and began looking in windows. While doing so, a neighbor came over and inquired as to what she was doing. Rheta told her she was trying to find a house to buy. The neighbor told her that the owners had gotten a divorce, and someone could buy the house by picking up two past due payments and then assuming the loan. We bought it.

The house was beautiful inside and out. It had four bedrooms, two full bath-rooms, and a two-car garage. The fenced back yard was large and adjacent to an undeveloped field although within a year new houses were being built. As nice as it was, the house was not a ranch. We kept searching for a place with land.

One evening, Rheta came home and announced that one of her patients found us a place very near him in Springtown. We wasted no time driving the twenty-five miles northwest of Fort Worth. The owner of the ninety-six acres allowed us to walk all over the land. We came back and said we wanted to buy it.

The owner, a dairyman, said we would need six-thousand dollars down and pay six-percent on the note he agreed to carry. We went home, opened a lock box, and took out a stack of U.S. Savings Bonds. We counted out six-thousand dollars and headed for the office of his attorney. When we left, I was happy. What about Rheta? Was she happy? I am not sure about that, for you see, Rheta was a city girl. I think she only agreed because she loved me.

Rheta immediately went to work putting her touch and style on the old farm house that was built in the nineteen-forties. The house had central heat in summer and central air in the winter. It had single walls so there was no way to insulate it. In the living room, there was a Dearborn stove fired by butane. The bathroom had a small electric heater. The bedrooms had no heat. We slept under quilts.

At any rate, we were ranchers now and ranchers must have cattle. So, I wasted no time buying a dozen heifers that were almost ready to have their calves. They were full blood Brangus which is a cross of Brahman and Angus. I also bought a Brangus bull.

It certainly was an exciting two years while we lived there. Our German shepherd chewed the tails off our Brangus calves and almost killed our pig Petunia. Petunia was given to us by a man who grew a crop of cantaloupes on halves. That crazy pig turned out to be quite a pet. She followed Rheta around everywhere she went. I think she must have thought Rheta was her mother. After the dog lacerated her badly, we took her to the butcher though we wondered whether we could eat her. We could not. The meat tasted horrible. We did not know that one must first feed the pig corn for two weeks to purge the pig chow that had been Petunia's diet. It seems all the hormones and antibiotics in the pig chow had not been cleared out of her system.

The meat never went to waste though. Earlier, I had been given a Brittany spaniel female. I enjoyed quail hunting a lot and wanted a bird dog.

When I brought the dog home, Rheta said, "Don, that dog is pregnant."

"No," I replied. "She's just fat."

Not long after that, the "fat dog" had a litter of ten puppies. Once the pups got old enough to wean we fed Petunia to them and as quickly as we could, found homes for them. That was the last bird dog I ever accepted for free.

With calf tails gone and Petunia no longer with us, what was a rancher to do? A few months went by before that question was answered. "All ranch hands on deck," was the call, even the five-year old ones. Rheta began having labor pains with our second child, Carrie. We drove the twenty-five miles into Fort Worth and admitted her to Harris hospital. She gave birth within an hour or so of her admission. After seeing our newborn baby, five-year old Steede and I spent the night with our friends who lived on Forrest Park Boulevard.

The next morning, Steede and I drove back to the ranch. I was concerned about one of our heifers that was due at any time. When we got home, we began looking for the cows. They were grazing in a pasture closest to the house, all but one. Yep, it was as I expected. The heifer that was due was missing. So, we started looking. We finally found the poor thing lying near a tree-lined fence. She was unable to have the calf without assistance and I had been AWOL. She had only managed to push the calf as far as the shoulder, and then only after great exertion on her part. She had exhausted herself and had given up. The calf was dead, and buzzards had already eaten the eyes.

I grabbed hold of the calf's feet and began to pull. "I'll get you free of this and all will be well," I told the cow. Good luck with that. Any hope of pulling the calf soon vanished as I strained and strained. Not only had I lost a fine-looking calf, a bull calf as it turned out, but if I did not get her free of the calf I would lose her too. So, I started pulling once more and gave the five-year old ranch hand his first assignment.

"Son, grab hold of a foot and help me pull," I called out to Steede.

That morning, I hired and fired my first ranch hand.

"Daddy, no, that's yucky," was his reply.

After being unable to persuade my son to help, I decided to drive to our neighbor's house and get him to help. He obligingly drove his tractor over and we hooked it to the calf's legs. Out came the calf but at the expense of the cow's pelvis. It was broken. Using the tractor's hydraulics and brute force on our part, we managed to get the cow into a stall. We fashioned a sling of rope and canvas and hoisted her to her feet using a block and tackle.

After several days in the stall, I finally decided to move the cow outside and try coaxing her to stand. After many days of going out and turning the cow over to avoid the development of sores, she got them anyway. It became very apparent that there was only one solution. I must put the poor thing out of her misery. So, with a sad heart, I went to the house and got my gun. It was hard to do but I did it. She would never have stood again.

Oh, and the five-year-old rancher? Well, he went back to being a kindergartener.

Peace That Passes Understanding

"And the peace of God, which passes all understanding, shall keep your hearts and minds through Christ Jesus." Philippians 4:7

What kind of peace can pass all understanding and what does "all understanding" mean? First, let us look at what the word peace means. If you ask anyone what the word means to them, I believe you will, by a great percentage, get one of two definitions, both of which will agree with Webster. The individual will either say that the word peace means the absence of war, or, the word means moments of tranquility. When have we ever had a prolonged period when war was not present somewhere in the world. Never. Also, when has tranquility ruled completely? Again, I say never unless you live in a monastery isolated from humanity.

With respect to war, so long as we are on Earth, we will never be at peace. The bible promises that until Christ returns, there will be wars and rumors of wars. Jesus said it in Matthew 24:6. And what about tranquility? Can we ever achieve peace in that sense of the word? No, we can't until we are in Heaven. Then what is the meaning of peace of God and peace that passes all understanding? I think it means a total confidence to rest in Jesus Christ.

Ephesians 6:12-13 informs us that we do not battle against flesh and blood but, rather, our fight is with the unseen powers of darkness, in other words, Satan and his forces. Are we able to defeat those powers? No, we can't unless we appropriate the power of the indwelling Holy Spirit as believers in Jesus Christ.

The Apostle Paul, in his letter to the church at Ephesus, tells us to put on the whole armor of God. This includes the helmet of salvation (you are a believer) and the sword of the Spirit (the bible). Prayer, bible study, and the power of Jesus fighting with and for us will result in us being able to come through various trials and temptations in life, understanding that Jesus will never leave, nor forsake us. That knowledge and our ultimate victory against Satan will yield to us the peace that passes all understanding.

The Days of Our Years

"Our days may come to seventy years, or eighty, if our strength endures; yet the best of them are trouble and sorrow, for they quickly pass and we fly away." Psalm 90:10

Psalm, chapter ninety, and verse ten is, at first glance, quite startling to one having attained eighty-one years. But it need not be if that eighty-one-year old individual is in Christ Jesus. With each year that passes after eighty, I feel I can see the finish line. Paul wrote about it in his letter to the Philippian church.

Sound Systems Then and Now

My first recollection of a sound system was the old Grafonola that sat on our back porch. Although new ones were ornately built with a polished finish, ours was a second hand one that obviously had been placed outside. Perhaps the owner no longer wanted it, but it still worked quite well as I recall. These music boxes came in different styles. Some resembled a suitcase vinyl album player of later years. Others were French furniture pieces. Ours stood upright about four feet tall. It had a lid and a cabinet drawer for storing records and hiding the speaker.

As best I can tell, the Grafonola we had was manufactured by RCA circa nineteen-fifteen. The speaker sat just below the turntable and was covered by a cloth grill. The arm that contained the stylus and swung out over the record and lowered to play the record looked like a connection of pipes one would find under a kitchen sink. The stylus head jutted outward and upward appearing to be screwed into a one-inch pipe a foot or so long and then into a flared pipe that fastened to the base on which sat the turntable. The stylus, which we referred to as a needle looked like a tiny finishing nail. We had a small box full of them.

A crank was located on the side of the cabinet. It was used to tighten the spring that spun the record. The "needle" was lowered onto a thick record which was constructed of graphite. The records looked much like vinyl record albums of today but contained only one song. The sound moved through the stylus arm contraption and down into the speaker.

The whole sound box thing intrigued me. At four or five years old, I could often be found standing on a chair on our back porch, lowering that tubular gizmo onto one of my favorite records. As I recall, Jimmy Rogers was a favorite. I was probably thinking something like, "RCA, you done plum good."

Five Nails That Turned the World Upside Down

Nails come in all sizes and shapes. When I was a kid, my carpenter father would often send me to his tool box to get tools and nails and such. As I think about nails, I can't help wondering about the size of five special nails.

Four of the most special nails were used to nail Jesus to a cross. Surely, they were among the largest. Those four will forever rank as number one in importance.as I see it. Of course, we must not neglect to mention the nail-like thorns that pierced Jesus' head so that blood flowed down His face.

The piercing of His side with a sword also opened a wound from which blood flowed. The four Gospels, Matthew, Mark, Luke, and John all discuss the crucifixion of Jesus. His vicarious death, His burial, and His resurrection are why we can have peace with God through faith in what Jesus accomplished on the cross. He died a substitutionary death bearing all my sins past, present, and future so that if I believe on His name, I stand forgiven forever. I did believe and I do stand forgiven forever.

But what about that single fifth nail? What is important about it that it should rank second only to the four that nailed Jesus to the cross? To understand that nail we must first understand what was going on in the early Roman Catholic Church.

In the fifteenth century, a common practice of the Roman Catholic Church was that of selling indulgences. A sinner would go the priest to confess. The priest would then require a price from the wayward parishioner before he or she could be forgiven and absolved of the sin.

Martin Luther, born in Germany in fourteen-eighty-three, studied religion at the University of Wittenberg and obtained a doctorate in theology. After his graduation, Luther began lecturing in theology. He was a devout Catholic priest and looked deeply into the Scriptures which were unavailable to most until the printing press was invented.

Paul's letter to the Roman church was the means by which God enlightened Luther. The young priest came face to face with the biblical truth that when we believe on the name of the Lord Jesus Christ we are saved by grace alone through faith alone and not by works. With this newly acquired truth burning within him, Martin Luther sat down and wrote ninety-five statements that, after much debate which Luther encouraged, became known as the ninety-five theses. These Luther sent to Catholic leaders and then nailed a copy of them on the door of the Wittenberg church. Luther was excommunicated but those ninety-five statements formed the bases for the Great Reformation.

So then, the next time you pick up a nail, reflect upon the four that nailed Jesus to the cross. But don't focus so much on the physical pain the nails caused because many have suffered greater pain than that type. The indescribable and unimaginable pain that Christ suffered was of a different kind. His agony was from the fact that even though He had never sinned, He became sin for us so that if we believe on His name, then our sin debt was nailed to the cross with Him, never to be used in a condemnatory judgement against us. Rather, if we believed in Jesus and accept His free gift of eternal life, then we now have peace with God. We are no longer His enemies. We are now His friend. Thank you, Jesus for enduring the pain of bearing my sin on the cross. Thank you, Dr. Martin Luther, for using that fifth nail.

It Was Shelter

I have memories dating back as far as the early nineteen-forties and perhaps, in a few cases, nineteen-thirty-nine in which case I would have been three-years-old. It seems to me that when we were young it was hard to find time for remembering, given the circumstances of working and raising children. But now I do have time and the older I get, the more I find myself reflecting upon the past.

One of my fondest recollections is that of the old home place in Panola, Oklahoma. Dad had been a farmer in the years before I came along. He had built the house where he and his first wife welcomed my half-sister and, two years later, my half-brother into the world. When the children were two and four years old, Dad's first wife died of pneumonia She was only twenty-four years old. He later married my mother and they had six more girls and me, the last of nine. At some point before I came along, the house where they lived burned to the ground and Dad rebuilt on the same spot. That was the house I remember.

A well was located just a few feet from the back porch, but we were not allowed to get close to it because the ground was unstable, and rocks occasionally fell into the well from the rock wall. This continued until Dad finally filled the well up with rocks and dirt.

The house was cold in the winter since it had single walls with no insulation. But, we had a large coal burning stove in the living side of the house as well as a wood burning cook stove in the kitchen. The bedroom side of the house required two or three quilts for comfort in winter. In summer, of course, the entire house got unbearably warm on days when the temperature went above a hundred degrees. We often had to sprinkle the beds with water in order to sleep. Centipedes and tarantulas were commonly seen on the floors or even the walls and ceilings. In at least two instances, a non-poisonous snake was discovered. Most of my siblings grew up there and left home. Soon after the last child was grown and gone except for Nadine and I Dad bought a house that had five acres just across the field from the old home place. In the community of Panola, most folks were poor, but we didn't know it. I thought our old house was a mansion. It wasn't. In fact, well, it just wasn't. But it was shelter.

The Light of Christ

"I will lead the blind on a way they do not know, by paths they do not know I will guide them. I will turn darkness into light before them." Isaiah 42:16

When I got up this morning, I went into the kitchen and flipped on the light switch. Suddenly, what I couldn't see before, I saw. It's like that with Jesus Christ, the object of Christian worship. His light dispels the darkness that once hid peace and joy from us and kept us from enjoying the fullness of His love. In darkness, we were stumbling our way through life in the dim light that Satan used to guide us on a path that served him. We saw only those things that mean so little compared to true life that can come from following Jesus' light.

The switch that will allow you to see His light is belief. "For God so loved the world that He gave His only Son that whoever believes in Him shall not perish but have everlasting life" John 3:16 Believe Jesus. Accept the free gift of eternal life that He offers you. Don't try to do things to please Him enough to get to Heaven. Simple child-like faith is all it takes. "If you confess with your mouth that Jesus is Lord and believe in your heart that God raised Him from the dead, you will be saved." Romans 10:9

So, if you have not experienced the salvation of Jesus Christ, do it today. In belief, flip the switch and let His light set you on a path you have not known, a path that is straight, one that leads to life eternal.

Empty Buckets of The Hearty

From the start, I wish to give Rheta credit for the title of this thought. Each morning before she goes to work we have our time of bible reading and prayer. We've found it so often makes the difference between a good day and a bad one. She almost always prays for children who will go to school without anyone praying for them to have a successful day.

The first time I heard her do that I remembered the prophet Eli's prayer for his sons because, as he said to God, their prayers might not be heard because they had no regard for God. When I was involved in teaching emotionally disturbed children, I often dealt with kids who came to school with baggage from home that caused them to "act out" in class but the little buckets in their hearts were empty.

On the surface, it might seem to the untrained eye to be just another case of ill-behaved brats. However, a bit of gentle inquiry will uncover many events in those young lives that played a huge roll in their outlook on life and how they viewed their own worth. Many, we found, never heard words of encouragement from their parents (or grandparents, as the case often was). Instead, they heard criticisms that wrenched at the heart. "You are not getting supper tonight. Go to bed. I wish I had never given birth to you." Or "You are nothing but trouble to me. You'll never amount to anything. Get out of my sight."

If you tell a child such things on a regular basis, they will often become what you accuse them of being. Without help, these children will become adults who journey through life with empty heart buckets, always seeking in the wrong ways how they can fill them. Certainly, it is easier for most adults to deal with feelings of emptiness but underneath the façade lies an empty heart bucket. It's more pronounced for children because they have yet to acquire needed coping skills. That's why it is so important for us as believers to include such children in our prayers because, in most cases, there is nobody at home to send them off to school with a prayer and words of encouragement or to pray for their academic success. In your next morning prayer, I hope you will include mention of these boys and girls as you do or did pray for your own. God alone can fill those empty little buckets of the heart.

The Old Home-Place

It's interesting how some recollections remain within one's grasp and willing to be called up on a minute's notice no matter the number of years that have passed. Digging into the recesses of the mind can uncover memories that can be relived over and over again. Take the old home place for example. Just a mere thought of it brings a deluge of images to mind.

The house sat about a quarter of a mile off of Number Five Road. It was rectangular. Earlier in the life of the house the bedrooms were on the north with the living area on the south. The structure was typical of dwellings in rural areas of the nineteen-thirties and forties. I have few recollections of when the rooms were arranged that way. I suppose my main memory of the bedrooms being on the north side of the house is of the fires that sometimes burned along the San Bois Mountains a mile or two away. I would lay awake and watch the twinkling fires as they burned in various interesting forms. I love mathematics and I notice things of nature that leave sinusoidal patterns as the fires did. I cannot account for why. There were no homes in the mountains except for a hunting cabin or two so men would sometimes set the fires to eradicate ticks

Our house had a front porch that was enclosed by bannisters. Often in summer, the porch was where family, friends, and neighbors would congregate until well after dark. Houses had no air-conditioners, so the porches were well used. It was a wonderful place to enjoy watermelon or ice-cream until temperatures cooled down enough to go inside. As I've mentioned in other parts of this book, we children enjoyed playing in the large front yard until it got dark and then we caught fireflies. But that was in the summer when we didn't have to go to school the next day. Life for children back then was void of any great concerns as I remember.

I mentioned the gravel road named for the Number Five Mine. A vein of coal ran along a ridge to the south of our house. I often played near an old mine shaft that had been abandoned after the vein of coal petered out. Momma said, "don't you dare go into those mine shafts." So, I didn't. There is no telling how many rattlesnakes were in those old shafts. But I was raised to obey my parents and I'm sure glad I fought back temptations to enter and explore the underground world.

This is the same ridge where I could often be found enjoying God's creation. He did make the rocks and hills, you know. Near the top of the ridge there were rocks aplenty. Large flat ones provided a sun-warmed surface for me to lie on and soak up rays next to my dog Colley. I wonder how many squirrels I killed on that ridge after my Black Lab Coley treed them. How many swamp rabbits did I bring home after hunting along Big Fourche Moline creek through a hundred or so acres of bottomland we owned on the south side of those hills? It's the same crest where my youngest sister and I would stand with Momma and look out over flooded fields after prolonged and heavy rains. I remember the first time I saw the Pacific Ocean. I thought to myself, "This is not as big as our bottom land. Of course, the horizon was playing mind games with me because I was just a young boy and the pond on our place seemed very large to me.

Dad was a carpenter, so we always had lumber and nails lying around. I built a sled with two by six runners and would harness ole Snip and hook him up. Off we would go on Number Five Road just me and ole Snip. Sometimes I took a fishing pole with me and branched off of Number Five onto a road that led to what we called The Narrows. The road gave access to Big

Fourche Moline creek without having to climb the ridge. There, a small washed out dam, made a perfect fishing spot. I caught lots of fish there, mostly Sun Perch and Goggle-eyed Perch. Summers back then were wonderful no matter how hot because we got a full three months out of school. So, I fished a lot.

A Final Word

I have enjoyed writing this book, my first, and I hope you have enjoyed reading it. I, not only, hope it has informed in funny ways about what life was like for me as a child growing up in rural Oklahoma, but my prayer is that God will speak to you in some of the scriptures I included. I hope you will be encouraged by them and perhaps even changed eternally by the high call of God on your life. I pray the Lord will bless you.

Made in the USA
Columbia, SC
27 November 2018